KB206649

드립니다.

B R A I N
T R A I N I N G
S U D O K U

브레인 트레이닝
스 도 쿠
3 6 5

브레인 트레이닝

스도쿠

365 문제

365

스도쿠 동호회

푸른미디어

1. 스도쿠란?

최근 「스도쿠」가 전 세계를 열광시키고 있다. 스도쿠는 가로와
세로 각각 9칸, 총 81칸으로 이루어진 정사각형의 모든 가로와
세로의 칸에, 그리고 가로와 세로 각각 3칸씩 모두 9개의 칸으로
이루어진 9개의 작은 사각형 안에 1에서 9까지의 숫자들을 겹치
지 않게 적어 넣는 숫자 퍼즐이다.

어떤 이는 스도쿠를 퍼즐의 천재 샘 로이드가 만든 「탱그램 퍼
즐」보다도 재미있는 게임이라고 극찬했는데 그 말은 조금도 과
장된 것이 아니다. 게임 규칙이 매우 단순해 누구나 쉽게 도전할
수는 있지만 풀기가 만만치 않은 지능형 게임이라는 커다란 매
력을 가지고 있기 때문이다.

스도쿠는 스위스의 수학자 레몬하르트 오일러가 만든 「라틴 사
각형」이라는 퍼즐에서 생겨났다고 한다.

한동안 잊혀졌던 이 게임은 1970년대에 미국에서 잠시 소개되었
고, 1984년 일본의 한 퍼즐 회사가 「스도쿠」라는 브랜드로 판매
해 인기를 끌면서 세계 각국으로 퍼졌다. 그 후 우리나라에서는
권위 있는 주간지 『일요신문』의 퍼즐 란에 「넘버 플레이스」라는
이름으로 실렸는데 최근에 들어서서 갑자기 전 세계적인 스도쿠

열풍이 일어나고 있다.

스도쿠는 수리력이나 지식으로 푸는 퍼즐이 아니라 오직 논리에 의해서만 푸는 퍼즐이다. 따라서 어린이부터 노인까지 누구나 사고를 집중시키기만 하면 재미있게 즐기면서 할 수 있다. 집중력과 추리력이 좋아지면 결과적으로 지능 향상과 두뇌 계발에도 도움이 되니 누구에게나 권할 수 있는 게임이라고 생각한다.

게임을 계속하다가 보면 스도쿠 퍼즐을 빨리 풀 수 있는 핵심적인 전략이 세워질 것이다. 또한 그렇게 되어야 난이도가 높은 스도쿠 퍼즐도 해결할 수 있다.

2. 스도쿠의 구성

이 책에서 편의상 전체 퍼즐을 「표」로, 3×3의 작은 표는 「상자」로, 숫자를 채워야 하는 공간은 「칸」이라고 말한다.

3과 6이라는 말은 3번째 가로줄과 6번째 세로줄이 만나게 되는 칸을 가리킨다.

상자들의 번호는 그림과 같이 좌에서 우로, 그리고 위에서 아래의 순서로 매겨진다.

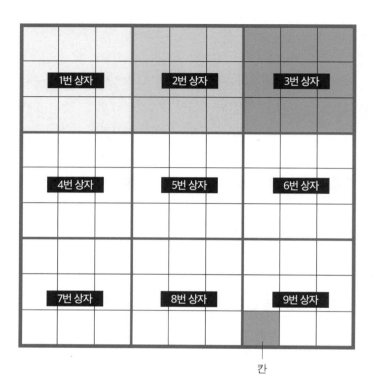

칸

3. 스도쿠를 푸는 기본 방법

일단 각 상자의 숫자들을 살펴 어느 칸이 비었는지 어떤 숫자가
빠졌는지를 가로줄과 세로줄에서 확인하라. 6페이지 퍼즐의 3번

상자를 보자. 이 상자에는 4가 없지만 7번째와 9번째의 세로줄에 4가 있기 때문에 4가 들어갈 수 있는 곳은 8번째 세로줄밖에 없다. 따라서 그 칸에 4를 기입하면 된다. 따라서 첫 번째 숫자와 두 번째 숫자 5를 해결하게 된다.

	1	2	3	4	5	6	7	8	9
1		1		4	6		2	3	8
2	6		4			2	7	9	◯
3	2		3	5	8		6	◯	1
4	7		6	9	1		5		
5	4	3						7	6
6			8		7	8	4		2
7	8		2		9	5	1		7
8		6	9	7			8		4
9	1	7	5		4	6		2	

4라는 숫자에 대해서 좀 더 생각하자. 5번 상자에도 4가 없다. 하지만 5번째와 6번째의 가로줄에 4가 있다. 들어갈 수 있는 곳은 4번째 가로줄뿐이다.

	1	2	3	4	5	6	7	8	9
1		1		4	6		2	3	8
2	6		4			2	7	9	5
3	2		3	5	8		6	4	1
4	7		6	9	1	◯	5		
5	**4**	3						7	6
6			8		7	8	**4**		2
7	8		2		9	5	1		7
8		6	9	7			8		4
9	1	7	5		4	6		2	

이번에는 1번 상자를 살피자. 보이는 바와 같이 2번 상자와 3번 상자에는 8이 있지만 1번 상자에는 8이 없다. 1번째와 3번째 가로줄의 8때문에 1번 상자에서 8이 때문에 1번 상자에서 8이 들어 갈 자리는 1번 가로줄과 2번 가로줄인데 1번 세로줄에 이미 8이 있기 때문에 8이 들어갈 칸은 한 군데만 남게 된다.

	1	2	3	4	5	6	7	8	9
1		1		4	6		2	3	**8**
2	6	○	4			2	7	9	5
3	2		3	5	**8**		6	4	1
4	7		6	9	1	4	5		
5	4	3						7	6
6			8		7	8	4		2
7	8		2		9	5	1		7
8		6	9	7			8		4
9	1	7	5		4	6		2	

2번 상자의 3번째 가로줄의 빈칸을 채우는 방법도 앞의 상황들과 비슷하다. 빈칸을 채울 수 있는 숫자는 7과 5인데 1번 상자의 빈칸은 1번째 세로줄에 7이 있기 때문에 7을 사용할 수 없다. 따라서 7은 3번째 세로줄의 빈칸을 채우게 되며 9는 1번 상자 3번째 가로줄의 빈칸으로 들어가게 된다.

	1	2	3	4	5	6	7	8	9
1		1		4	6		2	3	8
2	6	8	4			2	7	9	5
3	2	○	3	5	8	○	6	4	1
4	**7**		6	9	1	4	5		
5	4	3						7	6
6			8		7	8	4		2
7	8		2		9	5	1		7
8		6	9	7			8		4
9	1	**7**	5		4	6		2	

1번 상자에서 사용할 수 있는 숫자는 5와 7이 남게 되었다. 하지만 1번째 세로선의 빈칸은 1번째 세로줄에 있는 7 때문에 7을 사용하지 못한다. 따라서 5를 사용하게 되며 3번째 세로선의 빈칸에 7이 들어가게 된다. 또한 6번째 세로선의 빈칸에 자동적으로 9가 들어가게 되며 4번 상자 3번째 세로 줄의 숫자 1을 해결하게 된다.

	1	2	3	4	5	6	7	8	9
1	○	1	○	4	6	○	2	3	8
2	6	8	4			2	7	9	5
3	2	9	3	5	8	7	6	4	1
4	7		6	9	1	4	5		
5	4	3	○					7	6
6			8		7	8	4		2
7	8		2		9	5	1		7
8		6	9	7			8		4
9	1	7	5		4	6		2	

상황은 비슷하게 전개된다. 2번 상자에서 사용할 수 있는 숫자는 1과 3이 남게 되었다. 하지만 5번째 세로줄의 빈칸에는 5번 상자의 1 때문에 1이 들어갈 수 없다. 따라서 3이 들어가게 되면 1은 4번째 세로줄의 빈칸으로 들어가게 된다. 이제 나머지 숫자들을 해결하는 일은 훨씬 쉬워지게 되었다.

	1	2	3	4	5	6	7	8	9
1	5	1	7	4	6	9	2	3	8
2	6	8	4	○	○	2	7	9	5
3	2	9	3	5	8	7	6	4	1
4	7		6	9	1	4	5		
5	4	3	1					7	6
6			8		7	8	4		2
7	8		2		9	5	1		7
8		6	9	7			8		4
9	1	7	5		4	6		2	

7번 상자에서 사용할 수 있는 숫자는 3과 4가 남게 되었다. 여기서 는 그 결과가 매우 간단해진다. 1번 세로줄과 8번 가로줄에 각각 4 가 있기 때문에 숫자 4는 7번째 가로줄의 빈칸에 들어가게 되며 숫 자 3은 1번째 세로줄의 빈칸을 채우게 된다. 동시에 4번 상자 1번째 세로줄의 빈칸에는 남게 된 숫자 9가 들어가게 된다.

	1	2	3	4	5	6	7	8	9
1	5	1	7	4	6	9	2	3	8
2	6	8	4	1	3	2	7	9	5
3	2	9	3	5	8	7	6	4	1
4	7		6	9	1	4	5		
5	4	3	1					7	6
6	○		8		7	8	4		2
7	8	○	2		9	5	1		7
8	○	6	9	7			8		4
9	1	7	5		4	6		2	

	1	2	3	4	5	6	7	8	9
1	5	1	7	4	6	9	2	3	8
2	6	8	4	1	3	2	7	9	5
3	2	9	3	5	8	7	6	4	1
4	7	○	6	9	1	4	**5**		
5	4	**3**	1				○	7	6
6	9	○	8		7	8	4		2
7	8	4	2		9	5	1		7
8	3	6	9	7			8		4
9	1	7	5		4	6	○	2	

4번 상자에서 사용할 수 있는 숫자는 2와 5인데 여기서도 결과는 마찬가지다. 4번째 가로줄에 5가 있기 때문에 같은 가로줄의 빈 칸에는 2가 들어가게 되며, 5는 6번째 가로줄의 빈칸에 들어가게 된다. 7번 세로줄의 경우도 그렇다. 사용할 수 있는 숫자는 3과 9 인데 3은 5번째 가로선의 3때문에 9번 상자 7번 세로선의 빈칸으

로 들어가야 하며 숫자 9는 6번 상자의 7번 세로선에 있는 빈칸을 채우게 된다. 문제는 거의 다 풀렸다. 이제부터는 스스로 문제를 풀어보자.

브레인 트레이닝

스 도 쿠

3 6 5

QUESTION

SUD 365 OKU

DATE: TIME:

Question 1

8			5		6			3
	3							4
					1			
		8	7		4			1
1				9			7	
			8			4		
	8				2			
4		9		5			8	2
	6				8		1	

SUD 365 OKU

Question 2

	3	9					2	
8			2		3			1
			4		1			
	9	2				4	1	
		6	2					9
	7	4					6	
2				8				5
4		9		6				7
	8	1				6	4	

SUD 365 OKU

Question 3

1		3		8		5		
7					5		3	2
2			1		6			8
	1					4		
		2	3		4			
		5		2		3	8	
4			6	5	2			7
	7		9			2		3
5		1					4	

SUD 365 OKU

DATE: TIME:

Question 4

6		5					7	1
	8		6	1			3	
3				8	4			6
	1	8				2	9	
				7				
	7	6					4	
8			5		2			3
9			4		1			2
	2	3				4	5	

23

DATE: _____ TIME: _____

Question 5

	9		6				8	
		4					6	7
	6		2				9	
			3		4	9		2
7		8	5		1			
	3				2		1	
9	7					6		
	5				9		4	

SUD 365 OKU

Question 6

	5	2				1		
	4			9			6	
6					1			4
		4	2		7			3
	3			8			1	
1			4		5	6		
5			1					2
	2			7			3	
		7			4	5		

DATE: TIME:

Question 7

	5			7			8	
1			4		8			2
	2			6			3	
	8			4			7	
3		2	7		1	8		4
	6						5	
				2			1	
7		8	6		5			9
	1						4	

SUD 365 OKU

DATE: TIME:

Question 8

7			3		5			
		4	6			7	8	
1	6			7				3
6		5				4		1
	1			6			7	
			2		9	6		
2							6	4
	7				2	9		
		9	1	3				

27

DATE: TIME:

Question 9

8				7				5
	4		8				2	
6				4		7		8
	7				9			
1		3		6		8		7
			2				6	
2		7		9				4
	9				8		3	
3				2		5		9

SUDOKU 365

DATE: _____ TIME: _____

Question 10

	8		1		4			6
1		7						8
			8		3			
6	4			3			2	7
			2			9		
	9			6			4	
		6		7		5		
2						7		
			5		3		9	

SUD365OKU

DATE: TIME:

Question 11

			7		8			
6				4				8
	8		2		6			
		5				8		4
	4			6			1	
1		8				3		5
	5		9		4		2	
4								
			6		7		5	

30

DATE: TIME:

Question 12

3		1	4					
	2			1		9	4	3
			3	9				
9	3				1		2	
		4		3		5		
			2		4		6	
				2				
4	7			6			1	5
		9	7			6		

DATE: TIME:

Question 13

				5	4	3		
	4		6					5
3		5		7		9		
2							1	
		7		8		5		2
	9					6		
		8	7	6		1		4
4	1				5		7	
		2	3					

SUD 365 OKU

Question 14

	1			8				7
			6			1		4
6		3		1		2		
	3		7		8		1	
		2				4		
	7		9		6			2
7		8				6		
2				7				3
	5		1				8	

SUD 365 OKU

DATE: TIME:

Question 15

8				4		3		
		6	5	8		4		2
4			3				6	
		4						
9	8			6			7	
					9			1
	3				4	5		6
5		9		2	6		8	
		1		5			4	

34

DATE: TIME:

Question 16

	1	2		4			5	
				1		3		
6	3				7		8	
		4	7	6			3	
1						8		2
	5			9	8			
		1	6				4	
	6		2		4			3
4				3		5		

DATE: TIME:

Question 17

		7		5		6		9
1	2					3		
		6		7			4	5
	1		8		5			
	3	9				8		2
			6		2			
5	7				6		2	8
	8			2		5	6	
2		1						4

DATE: TIME:

Question 18

	4				5			8
	9	8		7			1	
6			8					7
	6			1	3			
		4				7		
			6	5			8	
					8			
		6		2				1
9		1			7		6	

DATE: TIME:

Question 19

		2				9		
	6	1		9		2		3
4	9		5	2	1		8	
		9		7		6		
	3		4		6		9	
		4		3		1		
	1		7		3		6	
3		6		5				9
	2						1	

SUD🌑365🌑OKU

DATE: _____ TIME: _____

Question 20

	4		6			7		3
1		3		7				
	6	2		3	5		9	
					4			5
	5		3	2		1		
					6			8
		8		6	1	3		
4	3	9						6
			5	4		9		

DATE: TIME:

Question 21

		4		8				
		6	4					9
1				3			5	
	8				1	5		
5				2				1
		9	5				6	
	6			9				8
2					8	3		
	9			1			2	

SUD365OKU

DATE: TIME:

Question 22

				8	6			
		5				2		
7	6			2			5	1
5			6	7	2			
2		7		1		5		6
			8	4	5			2
1	2			5			9	4
		4				7		
			9	3				

DATE: _____ TIME: _____

Question 23

		7		4			2	
9			2		7			3
				6				
	7		4				6	
6		9				1		4
	2				8		7	
				1				
1			9		6			2
	4			3			8	

SUD365OKU

Question 24

	5	2				9	7	
	7		6		1		2	
9				2				3
	3			1			6	
		7	8		5	2		
	4			7			1	
4				6				
			7		9		4	2
7		1				3		6

SUD●365●OKU

DATE: TIME:

Question 25

		2			3	1		
7	5		6				3	
		3		4		2		7
9	2			3			8	
		6	8		9	7		
	3			7			1	4
		4		1		6		2
2	1				7		4	
		9	3				7	

DATE: _____ TIME: _____

Question 26

		3	5					
	4	1				8		
	7			1	2		4	5
1			7				6	
		6				7		
7					8			2
	3		1	9			8	
		5				2		4
8				3	4		1	

SUD●365OKU

Question 27

8				3	9	6		4
	4	3			2		9	
		1						
4				6				
2	7			4				5
		3						6
	5	4				1	3	
3			4	2				9

DATE: TIME:

Question 28

		3			7	1		
7		8	1		3	2		
		2			4			3
	1						8	
8	3			4			9	7
	7						3	
4		7			9			
	5		8	1		9		
		1			6			5

SUD●365●OKU

DATE: TIME:

Question 29

	6	2					3	9
			3	2				
7	3			9			2	
		9				2		
8		1			7			5
			9			3		
	2			6			8	3
9				4	2			
		7				4		2

SUD365OKU

DATE: TIME:

Question 30

7	4		9	6				1
					3			7
9							4	
	1		2		6	5		9
3				7				
		7	4		1		3	
	3					1		
6			3					2
1				2	9		6	

SUD○OKU
365

DATE: TIME:

Question 31

		3		1		6		
6	5		4				3	7
1				6				
	9						7	
		6				9		1
	3		2					8
3				7			2	
9	1				8			
		2		5		7	1	

SUD365OKU

DATE: TIME:

Question 32

	5		1	6				
		7				3		6
4				2				9
	8			1			7	
		1	4		5	6		
	7			3			5	
7			3					2
		9			7	1		
1			6					

SUD 365 OKU

DATE: TIME:

Question 33

		7			8		1	
	1	6		4		5		
			5				7	
				4				1
9				2				
4			1					2
				1			9	3
		9		5		8		
	8		9					

SUD●365●OKU

DATE: _____ TIME: _____

Question 34

	9		5		2	4		
				3			9	
	4	7				8		
	1				5			
		8				9		1
4			8				5	
		3				2		4
	6			5			1	
7				1	3			9

SUD●365●OKU

Question 35

	3			7	5		6	1
		9		1				
	4		9					8
2					6	4		
	9			4			3	
		7	5					2
	1				3		5	
9	2					1		
	8		6	9				4

DATE: TIME:

Question 36

	2			5			4	
5		3	8					7
				7			8	
			5		2		3	
1		2		8		6		5
	8		7		3			
	4			2			9	
		6			7	8		
2			4				5	6

SUD365OKU

Question 37

2		7					1	9
	1				9		7	
		3	4				2	5
	2		1	7		5		
3			5		4			
				3	2	7		
	3				1		4	
4		2				9		1
	9	1					6	7

DATE: TIME:

Question 38

			3	2	6			
6		1				8		3
	3		4		8		7	
		6				3		2
4			8		1		9	
3		8				1		7
	5		2		9		6	
9		4		8		2		
	6		1		3		5	

SUD 365 OKU

DATE:　　　　　　TIME:

Question 39

	2			6			3	
7			9		1			8
1		8				2		
	1			8			5	
9			5		7			1
	4			9			8	
8		3				6		2
6			8		2			
	7			5			1	

SUD 365 OKU

DATE: TIME:

Question 40

		5				8		
9			5	8	2			7
		1				6		
2				3				5
	3			9		2		
6								9
		8				1		
3			1		8	5		6
		6						

SUD 365 OKU

DATE: TIME:

Question 41

8		6				3	1	
	3			6				
			1		8			5
		5				7		
	6			1			9	
		4				1		
6			2		3			1
				8			3	
	1	3						2

SUD 365 OKU

DATE: TIME:

Question 42

			5		7			3
1	4						5	
5		8				9		7
			6		9			1
8		9		7		4		
			2		5			8
6		2				1		
9				5			6	
	7		3		1		8	

SUD365OKU

Question 43

	1			2		6		4
		2	3		1			
8				7			3	
	3			8				
		4	7		9	8		
6				5			1	
	8			9				7
			6			9		
9		1			7			6

DATE: TIME:

Question 44

	3		2			9		
	4			8			6	
		8				3		7
4		7	1					3
	1			4			7	
2					6	8		
		4	5					9
	7			6			5	
5					1	4		

SUDOKU 365

Question 45

	7	2		3		9	4	
3			6		8			5
		5		7		8	1	
5			3		6			9
	6			8			5	
9			5		4			8
	5	8		6		3		
4			7		2			1
		1		5		4	6	

SUD 365 OKU

Question 46

					9	4	7	1
	4	9	7	1				3
	3				5	6		
	1		8	5		7		9
6			4			1		5
		8		6	7		2	
7		5	6				4	
4				2		9	3	
	8	3	9					6

SUDOKU 365

Question 47

			3			2		1
	2			8			4	
4		1			6			
		7		9				4
	3		2		1		8	
1				3		7		
			7					5
	8			1			3	
6		2				9		8

SUD365OKU

DATE:　　　　　　　TIME:

Question 48

8		7		4			1	3
					7			
6	1	2		8		4		
		8	3	1				
5					4			8
		1	2	5				
1		4		3		6		5
	6				5			
2		5	8			3		

SUD**365**OKU

DATE: TIME:

Question 49

	1					8		7
4				7				6
	7		2		9			
1				2		5		
	6		5		8		3	
		4		3				2
		1			4	7		
7				6				1
	2		3			6		

DATE: TIME:

Question 50

9		3			7			8
6			5			2		7
		2	8	9		3	5	
	2			6	5		3	
		8				7		
	9		3	7			4	
	3	9		2		5		
2		1			9			
4			1			9	2	6

SUD 365 OKU

Question 51

		9		4		1		3
7			1		8	4		
	1						8	
	9			3			5	
5		3	8		9			1
	7			6			3	
	6						7	
3		7	6		5	2		4
		1		8		9		

DATE: TIME:

Question 52

				6		5		7
5			2			8		6
	8		1		7			
		7		2		4	9	
2			8		3			1
	1	6				3		
7			3		9		5	
		1			6			
	6			8				9

SUD 365 OKU

Question 53

	2		7				8	
					9			3
8		9	6			1		
	8			6		4		9
			1		4			
9		4		3			2	
2		1			5	6		
			8		6			
	3						7	4

DATE: TIME:

Question 54

6					2		3	
		5		4				2
			9		7			
2		9		7		3		
	4	8			3		6	
		7		2		9		5
			7		9			6
8				1		7		
7			2					1

SUD 365 OKU

Question 55

		9				7	4	
	7			5			9	
4			9		7			
		4	8		2	5		
	1						2	9
		2			3	4		
			5		9			
	4			3				2
9				7				5

Question 56

	5				7	4		
9		6		3				1
	4		9				8	
3						1		
	2			4			6	
		8						5
	6				4			
				6		5		9
8		1	5				2	

SUD 365 OKU

Question 57

	2		8		7		4	
3		4				5		6
	6						2	
5			9		6			8
				2				
6			7		4			5
	3						5	
7		6				4		1
	5		4		8		3	

SUD 365 OKU

Question 58

		5			3			
						5		
				4				5
	5	4				9	8	
		6				4		
1						2	5	
5			9	2	8			6
			5		1	8		

SUD 365 OKU

DATE: TIME:

Question 59

					7			
		9		6	2		1	
							2	
2	4	7		5			6	
				7				
				3				
	8							
6				8	3		9	1
9								

78

DATE: TIME:

Question 60

			1		5			
9	5				6		8	4
	1	7			3	6	2	5
7			5					2
5	3			2	9			
2			3	1		5		9
6			9	5	1			3
3	8	9	7			2		1
		5						

SUD○365○OKU

Question 61

		3	6					4
6	8						1	7
					7	9		
9			1	8				
	2	8				5	4	
4				5	2			3
		6	7					8
	1			6			9	
8			2			4		1

DATE: TIME:

Question 62

	4		5		6		7	
				9				
7						6		3
4				8				7
	1		2		3		4	
3				7				6
		5						
	7			6			8	
	3		4		5		6	

SUD●365OKU

Question 63

	3	6				1	4	
	9				5			7
		2		1			6	
3			2			6		
	4			9			7	
		1			6			5
	5			2		3		
2			9				8	6
		7			8			

DATE: TIME:

Question 64

			8		2	4		9
	5	2					3	
1	4			9				7
3						6		
		1		3			8	
4						9		
5			6		1	3		
	6			4				2
	1		9		3		6	

SUD365OKU

DATE: TIME:

Question 65

	8				4			9
3		2						5
	6		7		1		8	
		3		7		8		
8								3
		5		9		1		
	5		4		9		2	
9						5		1
2			3		6			8

84

SUD 365 OKU

DATE: TIME:

Question 66

3	9				6	1		4
		7		3				9
8	1		9				2	
9			4	2		7		
	3						4	
		4		1	8			3
	8				4		7	
1		9		5		4		
7		5	8				3	1

DATE: _____ TIME: _____

Question 67

	2	6		5			3	
7			2			6		4
	3	1				2		
				9			4	
3			1		7			9
	8			2				
		4				3	8	
9		7			8			6
	5			4		9	1	

DATE: _____   TIME: _____

Question 68

6						2		4
	1		5	4		6		
7		4	2	3		1		8
8		9		7			6	
4			6		1			9
	7			9		5		2
	6	1		8	9	4		5
	4	2		6	5		3	
5			3					6

SUD 365 OKU

Question 69

1		9		7	5			2
5			4	2		9		
	4						5	
		3			1			7
	9			8			1	
2			3			6		
	5						9	
		1		4	6			3
8			7	9		1		5

SUD 365 OKU

DATE: _____ TIME: _____

Question 70

5	2		7				4	
		7			6			1
3				5		2	8	
	8				3			2
		9				3		
7			5				9	
	7	5		1				3
8			6			4		
	1				5		7	6

SUD 365 OKU

DATE: _____ TIME: _____

Question 71

	8			1			4	
7			9	4		6		8
	9	3			6	5		
		1		7			5	
8	6		2		3		9	1
	5			9		8		
6		4	8			9	2	
		9		2	7			6
1	2			6			7	

SUD 365 OKU

Question 72

1						3		4
			7	1				
5		3					8	6
	5		6		2			
	4			7			3	
			5		4		2	
4	8					2		
9				2	6			1
	2	5					4	

DATE: _____ TIME: _____

Question 73

	3			4				7
6	8			2		1	9	
		4	6		9	2	8	
8								1
		5		3		7		
9								2
	9		4	6	1	3		5
	5	1					7	
3				5				8

SUD 365 OKU

DATE: _____ TIME: _____

Question 74

	1			4			9	
7		4				8		3
	3		1		7		2	
		9	3		8	1		
4				7				2
		3	4		2	6		
	2		6		1		3	
1		5				9		7
	4			8			5	

SUD○365○OKU

Question 75

2				3		8	1	
			6					4
3			1		8	9		
	1	8				4		2
6								1
		5	4				3	
1		7			9			
8					6		7	
	3	6		1			8	9

DATE: TIME:

Question 76

			6		1	5		4
6	5						1	
1		9		7		3		
	6			8				
8	9		5				2	7
				1			3	
5		6		9		4		1
	8						5	
2			3		4			6

SUD 365 OKU

DATE: _____ TIME: _____

Question 77

1		6				8	4	
	5		4		7			1
7		4		1		5		6
	8		1		4		7	
		9		3		4		
	6		8		5		3	
6		3		8		7		4
	4		2		6		8	
8		5				2	6	9

SUD 365 OKU

DATE: TIME:

Question 78

		9		6	5			8
2	1			9				
		6					4	7
8			4			1		
	9				6		3	
		3			1			9
	2		9			8		
9				1			7	2
	7		5			3		

SUD 365 OKU

DATE: TIME:

Question 79

	4			8		9	6	
9		7						1
	3	1		9		4	2	
			8		6			
3		8				6		2
			7		3			
	1	3		4		2	9	
	5					7		6
4	2			7			1	

SUD365OKU

DATE: _____   TIME: _____

Question 80

5		3			7			
	4		1	6		2		9
		2			4			
	1						8	
8				4			9	7
	7						3	
4		7			9			6
		6	8	1		9		
		1			6		2	

99

DATE: TIME:

Question 81

		1	8		3	6		
	2	9				4	7	
	3			9			8	
3			6		8			2
				3				
1			4		9			7
	5			6			9	
	4	7				5	1	
	1		9		5		2	

SUD 365 OKU

DATE: TIME:

Question 82

4		6			8			1
1				4			9	
		3						
		9			1			8
	1			3			5	
2			9			4		
		1			5	6		2
7	2			8				
	6		2				7	

SUD 365 OKU

Question 83

6		1				4		2
2	8			6			1	3
			2		5			
5		6		8		3		1
	1		6		3		7	
		2		4		5		
8			3		4			5
1	5			2			4	9
		9				6		

Question 84

		9	1			7		
7				2			9	1
		5				3		6
9					3			
	7			9			3	
			7		2			4
8		7				2		
5	1			7				3
		3	8			1		7

DATE: _____ TIME: _____

Question 85

	8				9	7		2
6				7			4	
		1	8					6
	3				8			
8		9				3		
			5				8	
3		7			5	8		1
	4			3				7
		6	9					3

DATE: TIME:

Question 86

	4	7	9		5			2
2				8			3	
6			2	3				5
	2				6			7
9		3				4		6
		8					2	
8				4				9
	5			6			1	
1		9			8	5		3

DATE: TIME:

Question 87

3					8	7		5
	2			6			1	
6		1	7					2
		8	6					9
	6			2			7	
2					1	6		
7					2	1		6
	1			7			2	
8		2	1					7

SUD 365 OKU

DATE: TIME:

Question 88

9			8		4		1	5
		8		5			3	
1	5					2		
		9			2			6
	7			3			4	
3			1					
		5				4		3
	8			4				
4	2		7			1		8

SUD 365 OKU

DATE: _____ TIME: _____

Question 89

	6			2			9	5
		7		4			3	
2					3			1
9						4		
	3		4		2		6	
		6	5					3
7			2					9
	1			3		6		
6	8			7				4

108

SUD365OKU

Question 90

	5						2	
3			4		7			1
		1		2		3		
6				1			7	
		5	8		4	9		
	2			5				6
		7		9		6		
5			3		2			7
	8						1	

DATE: _____ TIME: _____

Question 91

9	3			1	8			5	4
		6			2	1			
		4					3		
2			8	7			9		
7								2	
	1			9	3			5	
	4								
3									
6				2	5		4	1	

SUD 365 OKU

DATE: TIME:

Question 92

6		4				1		5
	3		1		5			
5			6			7		3
	6		2				7	
		5		8				
	8				6	4	5	
3		6			7			9
			5		9		2	
9		1				8		7

SUD 365 OKU

Question 93

9				6			3	7
5		4						6
		6		4	9	8		
3		5						4
			2		7			
2						6		3
	6		3	5		7		
						2	6	
1	2				6			5

SUD 365 OKU

Question 94

1	3						9	8
	5		2		4		3	
		9		1		5		
5	2			8			6	1
7	1			5			4	2
		4	3		1	8		
		1	9		2	7		
3			1		6		8	4
	4	7		3		6		

SUD 365 OKU

DATE: _____ TIME: _____

Question 95

4				3		6		
	3				4		2	
		6	1					8
8			7			5		
	1			2			3	
		5			1			7
9						4		
	2		8				1	
1				5				6

DATE: TIME:

Question 96

		4				6	3	
7				6		8	2	
	6		7	8				
		9	6			4		2
	7							
4		3			9	7		8
				4	1		7	
	4	7		9				
6	5					1		4

SUD 365 OKU

Question 97

	7	5				1		2
	1				2		3	
4				1				6
	3		2		7		8	
2		6				3		
	9		1		4		2	
1				5				9
		4			6		5	
8		7				4		3

SUDOKU 365

DATE: TIME:

Question 98

3		7	9					8
	8					7	6	
	9				8			5
		6		3				4
			8	5	6			
5				1		3		
4			1				9	
	3	2					8	
8					5	6		1

DATE: _____ TIME: _____

Question 99

1		4	6				3		7
	8			3				1	
7		6	1		9	8			5
5		7		6		1			
	9		7	8	4		5		
		3		5		6			4
4		9	8		3	5			2
	1			7			4		
2		8			5	7			1

DATE: TIME:

Question 100

7		1	4			9	5	
	8			9				3
	5				1			6
		7	2			5	1	
3								4
	9	8			4	7		
1			7				2	
2				6			8	
	3	4			5	6		7

DATE: _____ TIME: _____

Question 101

			2	5		9	4	
8					6		7	1
5		1						2
	1			4	9			
9			1		3			4
			6	2			1	
1						2		3
	4		3		2			
2	3		5					7

SUD 365 OKU

Question 102

		9	2			6	4	
1	6		9		5			8
		4		6		1		
9	3				6		5	
		1				3		
	7			3			1	9
		5		1		9		
6			5		3		8	
4	9				8			1

SUD 365 OKU

Question 103

6							9	
5		3	9					1
		7	3	1		2	4	
	9			8				6
	4		2		5		7	
8				7			3	
	7	8		9	2	1		
	6				4	9		
2	3						8	4

Question 104

8	1					9		4
5				9		8		
	6	2		1			7	
			6	8	1			
6	3						9	5
			5	3	9			
	4			6		7	3	
3		9		4				
		6					5	1

SUD 365 OKU

Question 105

	3	9				2	7	
			4	3	7			
5		8				6		3
	6			5			3	
	9		2		1		5	
	2			6			1	
3		7				1		4
			7	2	3			
9		2				3		7

DATE: _____ TIME: _____

Question 106

9			5		7			1
		4				2		7
	5			6			3	
7			2		4			6
		2				8		
1			3		9			2
5	3			2			7	
		8				4		3
			7		8		2	

SUDOKU 365

DATE: TIME:

Question 107

6		8			9	1		5
		3	1			4		
	5			6			8	
2								1
		5		1		6		
3								2
	3			5			1	
		4	9			8		
1		2			3	5		4

SUD 365 OKU

DATE: TIME:

Question 108

9							4	
4			9	1		6		
		1	4				9	
5				7	2			
1						2		9
		2	1					4
	4		2		1	9		5
6		8	5		9			
	5					1		3

SUD⬤OKU
365

Question **109**

2					8		9	
	1	8		6				7
	6				2	8		
		6	9			5		2
		6				1		
5		9		8	1			
		3	8				4	
7				1		9	3	
	8		9					5

SUD 365 OKU

DATE: TIME:

Question 110

1		9				5		6
	2			9				
			1		2	3		
	4							3
		7		3		1		
8							9	
		1	5		8			
9				2			6	5
	5	4				2		

SUDOKU 365

DATE: TIME:

Question 111

	7			6			3	
5	3			1			6	8
			4		2			
		9	2		1	8		
8	1						9	4
		7	3		8	5		
			9		6			
3	9			7			2	1
	6			8			4	

SUD🄌365🄍OKU

Question 112

	3		5				8	
6		1		2			7	
4		7	3		8			9
	1					7	9	
7		6				5		1
	2	4					3	
			8		2	4		7
	4			1		9		8
1		8			9			

SUD 365 OKU

Question 113

	3	9				2		1
			4	3	7			5
5		8				6		3
	6			5			3	
	9		2		1		5	
	2			6			1	
3		7						4
			7	2	3			
	8	2				3	6	

SUD 365 OKU

Question 114

4			9		6		2	
	2			5		4		8
		8		4		5		
2					7			3
	6		4			2	7	
8			3					6
		1		7		9		
6				3			8	
	4	3	8		5			2

SUD 365 OKU

Question 115

	7	8	1	2				5
4		5						1
				7			8	
		4	5		1			2
1				8				6
2			3		6	8		
	2		7					8
7	4					9		
		6		1	3		5	

SUD 365 OKU

DATE: _____ TIME: _____

Question 116

8			2			5		
		9			7			1
	6							3
9					3			
		1				7		
			9					8
1				9			4	
	7					2		
		5			8		6	

SUD●365●OKU

DATE: _____ TIME: _____

Question 117

	5			9				4
6						8		3
4			3		6			
		4						7
						3		
		2						
1	4			2			9	
		9		4				1
3							4	

SUD 365 OKU

DATE: _____ TIME: _____

Question 118

	3						8	
1	9				3	7		
				6				3
		4	3				6	
	5			9			3	
	6				5	1		
8				5				
3		1	6					4
				3		2	9	

DATE: TIME:

Question 119

			3	8				
	5	7			9	4		
4							9	2
	3				1	6		
	1			7			5	
		4	2					3
1								9
		3	4			7	8	
7				5	6			

DATE: _____ TIME: _____

Question 120

	9	3	2			5	8	
					5			
6					8			4
	1	9						6
7						1	2	
			4					9
5			8					2
	6	4			3		1	

SUD 365 OKU

DATE: _____ TIME: _____

Question 121

	8				3	9		
9			8				5	
		2		7				6
	4		7					1
		5		8		4		
3					4	8		
7				5		6		
	9				7			4
		4	6				7	

SUD 365 OKU

Question 122

		1			6	2		
	6			2			4	
7			3					5
		3				7		
	7			1			9	
		4				6		
9					5			3
			8					6
2		6				8		

SUD 365 OKU

DATE: TIME:

Question 123

9					6			4
	2						1	
		6	8			9		
		4		5		3		
1			2		7			
				6				1
		2			3	1		
	6						7	
5			9					2

DATE: _____ TIME: _____

Question 124

	3		1		4			2
	9	7		3		8	4	
7			6		1			9
		1		8		2		
4			3		5			6
	8	9		4		1	2	
5			9		2		3	

DATE: TIME:

Question 125

		2		7		6		
9								1
8			4		9			7
			1		7			9
	6	3				1	8	
4			6		3			
3			2		8			
	8						3	
		7		3		4		8

SUD 365 OKU

Question 126

2			4		8			
		5				1	6	
	1						7	
3		2						9
		8				4		
6					5			3
	2						3	
	6	3				5		
			1		7			2

145

SUD 365 OKU

DATE: TIME:

Question 127

9						2		8
	5		8		2			
8			1					5
	2			6		8	4	
			2		9			
	3	9		4		1	5	
5			9					1
		7			3		8	
		8		2				3

SUD 365 OKU

Question 128

	8				7	4		
	6			1			8	
		1		5				2
		8			2	7		
4								8
		9	4		1	3		
7				4				
	1			2			5	
		5	9					7

DATE: _____ TIME: _____

Question **129**

	1					9	2	
3					2			4
7			4	5				
	5					2		
		3		2		1		
		1					4	
6				8	5			7
		6					5	
	4	7						2

DATE: _____   TIME: _____

Question 130

1				4	6			5
		8		9			1	
5			7					3
	5				9		3	
		4				1		
	9		6				7	
	2			7	3			
	8					7		1
		7		5				6

149

SUD 365 OKU

Question 131

6			1					5
				4	8	7		
3		7		5				
			6				1	
1				8				3
	2				7		5	
	7	1						4
		5	3					
				9	5		8	

SUD 365 OKU

Question 132

8						7		6
	2			6				
		4			5			9
			4			3		
	1			2			6	
		9			7			
6			9			4		
				3			2	
3		7						5

SUD365OKU

DATE: _____ TIME: _____

Question 133

2		1			4			
				9				7
	3			6			2	9
		4			5			
1	2						6	8
			3			7		
	6			2			9	
9				7				
4			8			5		6

SUD 365 OKU

Question 134

6		9						
				2		6		1
1			6	5			8	
			7					5
		7	1			3		
8					5		2	
	1			3	6			2
2		5		7				
	7					5	1	

SUD 365 OKU

Question 135

4					3			1
		2		6				
	7		9		8		5	6
7		5		1				
	3		8		4		1	
				7		3		2
	8		1		5			
1				2		7		
		3	4				6	5

DATE: TIME:

Question 136

			2			7		
5		7		9				
	6			7		4		5
7								
	9	2				3	6	
								2
2		5		8			4	
				3		1		
		9			1			7

SUD365OKU

Question 137

9				7				8
		2	4			5		
	8				3			4
		7	8				6	
8				2				9
	5				6	3		
5			3				4	
		4			5			3
	2			8		9		

SUD 365 OKU

DATE: TIME:

Question 138

			6			3		
	2				5		4	
4		5		1		6		
	4		5		2			3
		8				9		
1			8		3		7	
		9		4		1		5
	3		1				8	
		6		3				

DATE: _____   TIME: _____

Question **139**

2					7			8
		6			9	2		
	8		5					6
	6						2	
8			1	7	3			4
		7					3	
	3			9			4	
		5	2			6		
1				8			5	

SUD 365 OKU

DATE: TIME:

Question 140

	7		2					8
		8		5				3
3			1		6		4	
		6				7		
	5						2	
4		3				8		
	2		3		8			
1				7		6		
8					4		5	2

SUD365OKU

DATE: TIME:

Question 141

	3		2		9		4	
		6		1		9		
		9				1		
3			7					9
	9						6	
7				9	3	4		
	7							1
9				7				4
		8	4		6		9	

Question 142

5		4				6		3
1			9		7			8
				6				
	7			5			9	
		9		8		3		
	1			2			8	
	5			4				6
4			5		8			
		2				5		9

SUD365OKU

Question 143

7	3	9						5
			5	7		1		
			3			8		
5			1				9	
4				8				2
	7				6			3
		8			7			
		6		2	3			
		7				5	4	6

DATE: TIME:

Question 144

1			9		2			6
		4					3	
9				6			7	
	9	1				3		
7				8				1
		8				5	6	
	1			3				5
	3					6		9
2			7		1			

SUD 365 OKU

Question 145

5								1
	6			7			8	
1			2		9			4
		3		4		8		
	7		5		2		6	
		8				9		
			3		6	2		
	3			9			4	
7		6						9

SUD 365 OKU

DATE: TIME:

Question 146

				9	1	2		
	4		5				3	
5			8					1
			7		5	8	1	
3		5						6
	2		6		4			
	3							7
					3		5	
7		2	1	8				4

SUD365OKU

DATE: TIME:

Question 147

	4		3					7
		8			6			1
		1		7			9	
	9		8					4
		6				8		
7					5		3	
	5			1		9		
3			4			7		
8					9		6	

SUD 365 OKU

Question 148

9				6				
						3	6	7
					4	1		9
					7		2	8
	7	4	1		9	6	5	
5	6		3					
3		9	5					
4	8	2						
				1				2

SUD 365 OKU

Question 149

	8			9			3	
9	1						8	2
		2		5				
		1		7		2		
2			4		3			1
		3		2		8		
			9		6			
1	2						9	5
	5			1			6	

DATE: _____ TIME: _____

Question 150

		8	4				3	
9		6	2					
					5		6	8
		1					5	7
5				4				
	4					8		
1	3		7					
					3	7		2
	6					1	5	

DATE: TIME:

Question 151

	9					6		8
6			9		8			
		4		7		2		
	6						8	
		5		6		3		
	3						7	
		1		2		8		
9			6		1			
7							2	1

SUD 365 OKU

DATE: TIME:

Question 152

9				5				
	3	8			9		4	
			8				6	2
	5			6		3		
4			3		2			1
		9		7			8	
	9				5			
	2		4			1	5	
5				8		6		

DATE: TIME:

Question **153**

1				5			3	
		8			3			6
	3				2		9	
9		5			1			3
	6			4			7	
		7						1
		9	4			7		
2			1			3		
	1			3		2		

DATE: TIME:

Question 154

	6	2			1			
			6	3			2	
3						7		6
	2	4			3			
			8			4	7	
6			1				5	
4		7						8
	1		7		8			
				6		2		7

DATE: TIME:

Question 155

		2	8		9	3		
	4			8			1	
	5						2	
9			2		5			8
				3				
6			1		4			7
	9						6	
	3			8			7	
		5	6		7	9		

DATE: TIME:

Question 156

5		4						3
			9		7			8
	3			6		9		
	7			5			9	
		9		8		3		
	1			2			8	
		7		4				6
4			5		8			7
3						5		

Question 157

5				8			3	
	3		9				5	
4			2			1		
		5			4	3	2	
8								1
	7	6	5	1				
		4			3			5
	2				1		9	
1				9				7

Question 158

		8	7			1	2	
		3			5			
	5			6				9
2								1
		5		1		6		
3								2
8				5			1	
	7					8		
		2	8		3	5		

DATE: TIME:

Question **159**

6		3	2					7
	1			5			4	
				3				8
		7			8			4
	8						1	
5			4			2		
2			8					5
	5			6			7	
		8			7	3		

Question 160

6			5			3		
		1			6			
		8		1				2
	1		9					3
	6			8			5	
9				3			2	
4				5		2		9
8			4			5		
		3			2			

SUD365OKU

DATE: _____ TIME: _____

Question 161

				6		7	1	
6			3		5			2
1		3		4				
	3				4		5	
5					1			9
	1		2				6	
		1		2		6		7
3			1		6			4
	6	9				1		

SUD 365 OKU

DATE: TIME:

Question 162

3			5		4			
	6						2	1
		8		1		5		
8					3			7
	4			5			9	
2			7			8		5
		2				6		
	5			7			8	3
9			2		6			

SUD 365 OKU

Question 163

			7	3				8
1		9			4	7		
	3						2	
		5					3	9
			5		7			
4	1			8		2		
	6				5		7	
		7	6			1		5
5		1	8					

DATE: _____ TIME: _____

Question 164

5			6			2		
		6		8				7
	2				3			9
1			4			7		
	6			2			3	
		2			7			1
4			2				7	
6				9		5		
	1				6		9	

SUD 365 OKU

Question 165

	2		5		4	6		
		4						1
5					3	4		
4				5	2			
	6			8		7	5	
7				3				2
		5	2				8	4
	1			4		2		
		7	8		1		6	

SUD365OKU

Question 166

	8		2		3			7
						1		4
5	6				4			
6		7						1
				4				
1						3		5
			1				7	
8		6		2				3
	1				6	4		

DATE: TIME:

Question 167

7				6				5
	5				3		8	
		2				1		
			8		4		2	
3		8				5		9
	9		5		7			
		6				3		
	3		7				5	
8				2				6

SUD●365●OKU

Question 168

		4		5			8	
5			3					
		2				7		9
	5			2			1	
3			6		5		4	
	4			3			9	
		5				4		
	9			1	3			
2			5			9		1

SUD365OKU

Question 169

		4		8		2		
	3				9		8	
8								4
	2			6			4	
			3		4			
	8			5			1	
2								3
	9		5				2	
		5		7		1		

DATE: _____ TIME: _____

Question 170

5		9				3		
			6		2			
2						5		4
	7						2	
		5		1				7
	9						5	
3		1		2		7		6
		8		6		9		

DATE: TIME:

Question 171

						3		
	2				5		4	
4		5		1		6		
		5						
		8				9		
1								4
		9		4		1		5
	3		1				8	
		6						

SUD○365OKU

Question 172

7				2			6	
	4						7	
		2			6			9
			2			4		
4								7
		5			9	6		
	2		1			9		
	3						2	
6				8				5

SUDOKU 365

DATE: _____ TIME: _____

Question 173

			4					
	7				1		4	
4		8				7		9
	9							6
		4		7		9		
		1				6		
	4		8				7	
9					4			3

DATE: TIME:

Question 174

	1				5			3
	7					5		
		4	7				1	
		6			3			8
			6					
9			2			6		
4					1		6	
		5						
7			8			2	5	

DATE: TIME:

Question 175

9								6
	2				3		9	
		8	4			1		
	5		6		1		3	
	9		5		2		1	
		9			4	5		
	6		7				4	
5								1

DATE: TIME:

Question 176

		8						5
9		6	2					
				5			6	8
		1						7
				4				
2						8		
1	3		7					
					3		1	
	6					5		3

SUD 365 OKU

DATE: TIME:

Question 177

7	6							
				2	7			6
4						7		3
	1			6				
	5		1		2		9	
				5			6	
		6				8		
5			7	3				2
	7						3	

SUD 365 OKU

Question 178

		6	7				8	
1		7						
					1		6	3
		8			9			
9				4			5	
			5			2		
7	6		3					
				7		3		
	1					6		7

SUD 365 OKU

DATE: TIME:

Question 179

5		7				4		1
				4				
6			9			3		
	5				4			
		9				5		
			3				6	
4					5			7
	2			1			8	
		8				9		

Question 180

		8		2				1
					4		9	
5		6				4		
2		7						
	5			3			1	
		3			8			2
3			8			7		4
	4	5						
			4	6		2		

DATE: TIME:

Question 181

1					2			
				1		2	6	
4		3				7		
7								9
	3			5			4	
5								8
		9				8		
	4			8			7	
	7		1				5	

SUDOKU 365

Question 182

	8						5	
	1		3		6		8	
	9						7	
	7		4		3		9	
	2		7		5		4	
	3						6	
	4		6		8		1	
	6						2	

DATE: TIME:

Question 183

		4		8		6		
	3			9			7	
		9				1		
2								6
			6		5			
9								7
		2				7		
	9			6			5	
		5		1		9		

SUD 365 OKU

Question 184

	4		1			2	3	
		3						
	6				7			5
4			2			1		
		7			8			6
			5				4	
2						7		3
	8	9			3			

SUD 365 OKU

Question 185

1			4		5			7
		6				1		
	5						3	
3				2		7		6
			5		3			
9		5		7				3
	9						2	
		8				6		
5			2		4			9

DATE: TIME:

Question 186

			1			2	4	
1						6		
5	2				4			
		5			6			
	9			4			8	
			5			7		
			3				6	4
		7						5
	6	3			8			

SUD365OKU

Question 187

	8					6		
				6		1		4
5	6				4			
6		7			2			
				4				
			6			3		5
4	9		1				7	6
				2				
		3					8	

DATE: TIME:

Question 188

2	4	7						1
					6			5
				2				3
6				3				
		9		1		2		
				5				4
5				9				
4			8					2
	7					5	1	

SUD 365 OKU

Question 189

9								7
	3						4	
2			1	9				5
		2				4		
		9		3		5		
		4				6		
8				2	6		1	4
4							3	
	7							

Question 190

					9			5
	4						8	
8		3	1				6	
			7			5		
	6						9	
		1		4				7
	3			1				4
	9						5	
		5	8				1	

SUD 365 OKU

Question 191

8					1			6
	4		2				5	
		5				1		
	1						6	
6				2				1
	3						7	
		1				7		
	2				5		4	
9			7					3

SUD 365 OKU

Question 192

			1			4		
	4				9		1	
3				4				
	1				6			4
		4				8		
2			8				7	
		5		8				7
	3		7				6	
					2			

DATE: TIME:

Question 193

			9		5			
2				8				4
6								5
	2				6	3		
		3					5	
	7		8			9		
8								9
7				6				8
			7		8			

DATE: TIME:

Question 194

			5		6			
5				9				8
		1				6		
4								7
	1			5			4	
3								6
		5				2		
2				6				5
			4		5			

SUD●OKU 365

DATE: _____   TIME: _____

Question 195

		7			6			1
	8			4			2	
6			9			3		
	5				4			
		9				5		
			3				6	
		6			5			7
	2			1			8	
3			7			9		

DATE: TIME:

Question 196

6			2		5			9
	9						2	
		5				6		1
1		4		5		8		2
		6				1		3
	1						9	
5			6		1			7

SUD365OKU

DATE: TIME:

Question 197

		9			1			7
	2			3			1	
8			5			6		
	9				4			
	1						9	
			9				6	
		3			5			4
	5			4			8	
4			6			3		

SUD365OKU

DATE: TIME:

Question 198

	7				9		3	
2					8			5
		6	5			1		
	2					4		9
				9				
9		4					8	
		1			6	9		
3			9					7
	8		1				2	

3					4			
			8			9		
4		6				8		3
	8		2		1			
9							7	
			4		3			9
		1				3		5
	3		5			6		
	6				8			

SUD 365 OKU

Question 200

1								2
		8				5		
	6		7		9			8
		2				7		
4				1				6
		3				8		
	8		4		2		9	
		9				3		
5								1

SUD 365 OKU

DATE: TIME:

Question 201

	7						6	
		4	2		3		5	
	9				7			4
	3		7					
9		8				6		2
				6			8	
2			1				3	
	5		8		2	7		
4							1	

DATE: TIME:

Question 202

			2		1			
	4						3	
		3		5		9		
7			5					9
		5				6		
8					6			2
		4		9		3		
	2						1	
			8		7			

SUDOKU 365

DATE: TIME:

Question 203

			7	9		2	3	
3					1			
7					3			
	5							7
9	4			5			6	3
8							2	
			9					2
		8	5					9
		6		1	2			

SUD365OKU

Question 204

	4	5						
			9	4	7			
			3			8	6	
8					6	4		
	6						1	
		7	1					9
	9	1						
			5	8	3			
						2	3	

DATE: TIME:

Question 205

			8		3			
	4							
9		2				4	1	8
		8	3		7	2		
		4	2		9	8		
6		3				1	8	2
	8							
			1		6			

SUD 365 OKU

DATE: TIME:

Question 206

2					3	1		
		3	9				8	
7					5			3
1						8		5
		4		9				
	3						9	
	7			8			6	
	2			7				
9					1	5		8

DATE: TIME:

Question 207

4				8				5
9			3					8
		1				9		
		7		4			9	
1			2		6			3
	8			7		4		
		9				2		
				5				
3	1			2			4	9

SUD365OKU

Question 208

2					3			1
1	4					3		
			9			6		
9								
6		4	2		5	8		7
								4
					4			
	7	9					2	8
			5			1		

SUD 365 OKU

Question 209

		7	9				8	5
2	5				4			
			3					1
	8					4		3
				5				
7		6					2	
9					3			4
	7		6					9
	1				7	3		

SUD365OKU

DATE: TIME:

Question 210

4					5			1
		8		9				7
7	6		1				9	
3						7		
	2			5			6	
		1						4
	1				6		7	9
8		7		3				
			5				1	

SUD 365 OKU

Question 211

	6					8		
			9		8			6
5		3						4
	5				7		4	
	1		8				5	
		8				2		5
			6		9			8
7		6						

Question 212

8			6					2
1				3				6
	4						9	
4		3						5
				5				
6						9		1
	8	9					2	
				6				9
7					3			4

SUD365OKU

Question 213

1	5			8			4	6
3								8
			3		1			
		7				6		
4				5				2
		1				4		
			2		6			
7								1
9	4			7			6	3

SUD 365 OKU

Question 214

	6				7			5
2			8			9		
		4		5			6	
	9		2					8
		5				3		
3					4		5	
	1			6				
		7			8			4
5			9				7	3

DATE: TIME:

Question 215

				5				
	4			7			6	
		6	4		8	2		
		2				6		
1	6			3			9	2
		3				8		
		7	3		5	9		
	8			4			3	
				9				

234

DATE: TIME:

Question 216

5	8		7					9
			3					2
2					9			
7			2					4
	1						2	
		8			5	9		
			8					1
	3	7			4			
8					6		3	7

SUDOKU 365

Question 217

	9		7				8	
		4				9		
	7			1			3	
5			8		7			
		6				2		
			6		2			4
	3			7			9	
		1				7		
	6				9		2	

SUD🌑OKU

DATE: TIME:

Question 218

3		7	9				2	
	9				8	1		5
		6		3				4
				5				
5				1		3		
4		5	1				9	
	7				5	6		1

DATE: _____ TIME: _____

Question 219

1	8							
	2				8	1	6	
				7		3		
	6		1		2			
		8				5		
			4		6		9	
				3				
		4	5				2	
	7	2					5	4

Question 220

	6			4				2
3							8	
	1		8		2			7
		2				7		
5				2				1
		4				2		
			4		8			
8								4
9	4			5			2	3

Question 221

			6			7	2	
	7	8			2			
2				7				4
			8					
	5	4				2		
	6				9	8	4	
5				2			8	
	9		5			6		7
		7			4			

SUDOKU 365

DATE: _____ TIME: _____

Question 222

4		1				2		8
	6						4	
7			3		8			5
		3	6			7		
				1				
		4			9	6		
9			7		5			3
2		6				8		7

DATE: _____   TIME: _____

Question 223

1	7						4	6
5			4					1
			9					
			4			6		
		4	9		5	1		
	1		3					
			5					
3				7				8
6	5						3	4

SUD 365 OKU

DATE: _____   TIME: _____

Question 224

		4		2		3		
	5						8	
3				6				5
			8				7	
2		8				5		1
	3			9				
1				8				9
	2						4	
		5		9		6		

SUD 365 OKU

DATE: TIME:

Question 225

3		1	9					7
	7			1			2	
			3					9
			6		7	4		1
	1						3	
6		4	1		5			
8					9			
	5			4			9	
2					1	6		3

DATE: TIME:

Question 226

			7		2			
	1	2				5	8	
	5			3			9	
7				6				2
			3		7			
1				9				3
	4			2			3	
	2	6				9	1	
			6		1			

DATE: TIME:

Question 227

	1						5	
7		9		3				4
			7		4	1		
		6				8		
	3			7			9	
		2				7		
	2		8		3			
1				4			3	8
3								2

Question 228

	2					9		6
9			2		3			
		4		6				
	3			7			8	
		7				3		
	8			4			9	
3				2		1		
5			4		7			9
7						5		

DATE: _____ TIME: _____

Question **229**

6								8
	2						7	
7			5		9			2
		4		3		5		
1			9		8			3
		2		1		4		
		9	4		1	3		
3		6				7		1

Question 230

		7	8		5	4		
	6						3	
5				9				2
4								1
		1		5		3		
2								5
3				7				6
	4						7	
		5	6		9	8		

DATE: _____ TIME: _____

Question 231

2			7					8
					3	6		
	7	8						3
7			8			3	6	
8			6		5			7
	2	3			7			5
3						1	8	
	9		1					
		4			6		5	

SUD 365 OKU

DATE: TIME:

Question 232

	8	3						
	7	2	4	3		8	9	
						4	3	
			2		7			
	4			5			6	
	3		6		9		1	
	2	4						
	1	8		7	4	5	2	
						6	7	

SUD 365 OKU

DATE: TIME:

Question 233

2				7			1	
		7		8			2	
	9		4					6
		6			2			9
					8			
9			1				3	
8				1				
		9			7		4	
	2			5		7		1

SUD 365 OKU

Question 234

4						1		9
	3		9		5		7	
5				7				3
			3		9			
6	1						3	7
			7		6			
				6				5
	6		4		8		1	
3		2						8

DATE: _____ TIME: _____

Question 235

			5		7			1
6		4				2		
	5			6			3	
7			2		4			6
		2				8		
1			3		9			2
	3			2				8
	7					4		
			7		8			9

DATE: TIME:

Question 236

								2
8	3			5			4	
	1		2				6	
		5			3			
3	7						2	6
			1			3		
	6				4		9	
				6				3
1		8				6		

Question 237

	6		1		5			
	8			3			2	
		7		2		1		
4								3
	7	9		5		6	8	
8							4	
		3		6		4		
	2			8				
1			7			2		6

SUD(365)OKU

DATE: TIME:

Question 238

9	7						8	
4				3				5
			8	1	9			
		6	9		1	5		
	9	8		5		2	6	
		4	6		2	9		
			1	4	5			
3				6				9
	4						5	1

SUD365OKU

DATE: _____ TIME: _____

Question 239

	7					3		8
5			4					
				1				2
	8				5			
		4		6		7		
			9				2	
3				8				
					2			5
9		7					4	

SUD 365 OKU

DATE: TIME:

Question 240

	4				5	8		
2		3		1			7	
			2					5
		6		4			8	
5	9		6		7	4		
		4		5			9	
			4					3
8		2		7			5	
	3				6	2		

Question 241

				7				
	3		9		6		5	
	2	6				8		
		4	1			3		
9				8				7
		2			3	1		
		5				4	3	
	4		2		5		7	
				1				

DATE: TIME:

Question 242

2		5				6		4
	1					5	2	
				1				
			7		6			
	8		5		9		7	
			1		8			
			9					
	7	8					6	
3		4				2		7

DATE: TIME:

Question 243

		2		6				
8								2
	3				1		5	
	4		1					
		9		8		1		
					6		3	
	5		7					
6					2			1
						8		

DATE: _____ TIME: _____

Question 244

					7		1	
4								
		9	2				8	
6				4		5		
				7				
		8		1				6
	4				5	1		
								2
	5		8					

SUD365OKU

Question 245

		2	1					
						4		
	4		6					5
						3		2
			9	1				
7		5						
8					5		6	
		3						
					1	2		

DATE: TIME:

Question 246

			2			5		
8		7	6			2		
				5		7		6
			2					
5		3						
		1				3		7
					6			
		5			2			

DATE: _____ TIME: _____

Question 247

	9					8		7
				6				
5			1					
		4			5			9
	8			7			6	
7			9			3		
					8			4
	1			3				
		7					5	

Question 248

9					3		6	8
			2	5				
	5						9	
			4					1
		8		9		3		
1					8			
	4	9					3	
				8	5			
2			1					9

SUD 365 OKU

Question 249

					7	4		
	6			1				9
				5				
					2	7		
4								8
		9	4					
7				4				
				2			5	
		5	9					

DATE: TIME:

Question 250

							9	6
8	1							
				2		7		
3		2	4		1			
						1		
			7		6	2		
		9		1				
							6	3
6	4							

DATE: TIME:

Question **251**

	6		3		5		7	
8								2
			4					
		8				7		3
				7				
1		7				6		
					7			
3								9
	5		2		6		1	

Question 252

			8		9	1		
		3					8	
	2			5				
		4		1				3
9			7		5			8
6				2		4		
				3			7	
	1					6		
		5	4		6			

DATE: TIME:

Question 253

2						5	1	
					3			4
		8			1			
7						3	6	
			6		5			
	2	3						5
			9			1	8	
			1					
	8	4						

Question 254

		2	1					
	8					4		6
	4		6			1		
							8	2
			1					
7	6							
		4			5		6	
1		3					4	
					1	2		

Question **255**

			7			3		
9		3	8			2		
						6		9
			1					
2		8						
		2			8	9		5
		5			4			

DATE: TIME:

Question 256

	6	3				5		
	2		7		3		8	
		5			9			
		7		3		6		
			5			2		
	8		1		5		4	
		2				1	6	

SUD 365 OKU

DATE: TIME:

Question 257

		3	9		6	1		
			4		2			
		8				3		
7	9						3	6
4	3						2	8
		4				2		
		5		9				
		9	6		1	4		

DATE: _____ TIME: _____

Question 258

9			5					
	3	8		9			1	
							7	
			3	7		1		
		6				5		
		1		2	8			
	9							1
	4			8		2	6	
7					3			

SUD365OKU

DATE: TIME:

Question 259

		1			4			
				9				7
	3			6			2	9
		4			5			
1	2						6	8
		3			7			
	6			2			9	
9				7				
			8			5		

SUD 365 OKU

DATE: _____ TIME: _____

Question 260

					7			
			9					
7				8				
	8	1		5				
	2		6			8		
	3				4		6	
			4			9		
		6		2			7	8
		9		7				6

DATE: TIME:

Question 261

			8			7		
					2			
		3	9	5				
6						1		
					9		5	
	8		5		1	4		6
				1				4
7	4		2					3

SUD365OKU

Question 262

	6			4			5	
		1	3	9				
2	1	9	4	6	5		7	
		2	7	8				
						5		
	5			1			8	
								3

281

DATE: TIME:

Question **263**

				4				
	3			9			6	
				6				
	5			3			2	
						6	3	1
3	8	5		1		9	7	4
						5	8	2

DATE: _____   TIME: _____

Question 264

	3							
	9			3			4	
	1							
						4	5	1
	5							
			3	4	9			
	4		7	2	5		1	
			6	8	1			

DATE: TIME:

Question **265**

	1							
	4			8			2	
	2							
	7			1			9	
4	6	2						
1	5	7		4		2	6	3
9	3	8						

SUD(365)OKU

DATE: _____   TIME: _____

Question 266

				4				
	1			7			8	
				6				
6	7	9						
4	5	8		2				
1	3	2					4	
	2			5		9	3	8

SUD365OKU

DATE: _____ TIME: _____

Question 267

			6	3	1			
	9		4	8	7		6	
			9	2	5			
	7					9	3	2
				9				
	1			4			9	
				7				

DATE: TIME:

Question 268

						3	9	6
9	3	2		6		1	4	5
						7	2	8
	6			5			8	
	1					8	3	9

Question **269**

							9	
	2			4			6	
							4	
						5	1	6
	4			6		8	7	9
						4	3	2
	7							
	6			3			2	
	3							

DATE: TIME:

Question 270

8	1	3						
						8	1	2
9	6	2		1				
4								
						1		
		5	3	9				
	1							
					4			

DATE: _____ TIME: _____

Question **271**

			6		4			2
			3		1			
	7		2		8			
	6			1				8
				6				
				2				
3								
						2	4	7
				8				

Question 272

						9	3	2
5				1				8
						1		7
		5						
				7		9		
4	2	7						
						3		5
	9		7			2		

DATE: TIME:

Question 273

5								
3			7			2		
4								
						6	4	8
		9	3	6	2			
					4	3	9	2
	3			5				
				8				
				2		7		

DATE: _____ TIME: _____

Question 274

		2						
		9				4		
		4						
			3					
9						3		
						6	8	2
	6					9	5	3
			6	2	3			

293

Question 275

	3			2			5	
				1				
							9	
4	5	8						
			9	3	8			
	4		2	5	7			
						2	3	7

DATE: TIME:

Question **276**

5						9		
			3			7		
						2		
		7						2
		6		3				
		5						
							9	
4	1	8		9				
6	5	9						

DATE: _____ TIME: _____

Question 277

							3	
		6	4	8				
	9							
5	6	8						
			6		5	8	1	
7	4	1						
						1		
		5						

SUDOKU

365

DATE:　　　　　　　　TIME:

Question 278

						9		
		1				2		
4					1	7		
1			7					
2								
		9				1	2	8
		7						
		1					7	
	2							

DATE: _____ TIME: _____

Question **279**

	4							
	7				5			
	6					8	1	5
			2					
			7					
3			9			4		
							4	
	5			9			8	
							2	

298

		5					1	
		8			5			
		3						
			7					1
5	8	2						9
								8
	4			2				
			9		8			
			8					

SUD 365 OKU

DATE: TIME:

Question 281

								8
	5						2	
					5	3		
5				1				
			8					
							7	
			6					4
	7						5	
	6					1		

SUD365OKU

DATE: TIME:

Question **282**

1				2				
	8					4	1	6
		6						
	9							
				5	1	6		
			8					2
	1						4	
						7		

DATE: _____ TIME: _____

Question **283**

		6						
		5				3		
		3	2					
						6	2	4
	3			9				
	8							
	1		6					8
				2				
					9			

Question **284**

1	2	7	6	9				8
							6	
						1		
	9							
			7					
2				1	9	4	8	7
		1						

DATE: TIME:

Question 285

	7								2
	5				1				
	8					3			
	4								
	9			4			3		
							7		
		4					8		
			7				9		
9							6		

DATE: _____ TIME: _____

Question 286

		7						6
		8						
					2			
					7		1	
7	4							
								7
			9					1
		9						
		4						

Question **287**

								8
	8							
				1				
						3		
	1							
						8	7	5
					3	2	6	7
2					4	9		
					7	5		

Question 288

		7		2			3	
	8	3			6	2		4
	5							
4				9				8
							2	
2		4	8			9	6	
	3			6		4		

DATE: TIME:

Question **289**

	8							
				6			1	
				2				
				9				
2	3	8				6	4	9
				8				
				4				
	6			1				
							9	

DATE: TIME:

Question 290

					8			
	3	8					2	
	7					1	6	
			9					
7								
								1
	4	3					7	
	2					8	3	
			8					

SUD 365 OKU

DATE: TIME:

Question 291

	9					8		7
				6			2	
5			1					
		4			5			
	8						6	
			9			3		
					8			4
	1			3				
2		7					5	

SUD365OKU

Question 292

		9	5					
		1	6					
						1	6	
						6	2	
			5					
	4	5						
	1	2						
					3	8		
					4	2		

DATE: TIME:

Question 293

		3	6					
		7	5					
						4	5	
						8	9	
			8					
8	3							
4	7							
					7	6		
					5	9		

Question 294

				3				9
	8						3	
						7		
				3				
2				5				1
			7					
		3						
	4						6	
6				7				

Question 295

				4				
	3						1	
		7						
			2					
8				5				9
				8				
						4		
	6						7	
				8				

DATE: TIME:

Question 296

	7	2	4	3	1	8	9	
	6						3	
	5						4	
	4		5				6	
	3						1	
	2						8	
	1	8	9	7	4	5	2	

DATE: TIME:

Question 297

7								2
		9	8		2	5		
		2	6		5	7		
		4	3		1	2		
		3	4		8	6		
9								3

Question 298

		7	6	1				4
				5				
				4				1
								9
3	2	9				8	7	5
6								
4				8				
				7				
8				3	9	2		

SUD365OKU

Question **299**

		8	4		1	3		
		3	8		9	1		
1				5				2
		4	1		6	2		
		1	5		3	6		

SUD 365 OKU

Question 300

			1					
	2					6		
						8		
		2						
7				8				4
		4	7		1			6
								9
	7		9		8		1	
	9					2		8

DATE: TIME:

Question 301

	7		6		8		1	3
	9	3						
	6				4			
	3	5			7			4
							7	
4				7			8	
		8					3	2

SUD 365 OKU

DATE: _____   TIME: _____

Question 302

					1	7	2	
5			9	6				
4								
7							8	
	5			7			1	
	3							4
								3
			4	5				1
	4	9	2					

SUD 365 OKU

DATE: TIME:

Question 303

				2				
	5			4			7	
							1	
	8			6			4	
	2							
	3			5			2	
				3				

Question **304**

	2			3	7		4	
	7							
	4			7			9	
							5	
	5		3	8			1	

SUD 365 OKU

Question 305

	9			2			1	
							9	
			5					
	2	7	4	8		3		
			1					
1								
5			3			7		

SUD365OKU

DATE: TIME:

Question 306

	2						7	
8				1				
							9	
	1							
			9					
							2	
	9							
			2					7
	8						1	

SUD365OKU

DATE: TIME:

Question 307

				5				
9			4			8		
		3						
1			7			2		
		3						8
3			2			6		
								3

DATE: _____ TIME: _____

Question 308

				8				
		7						8
4			5			3		
6			2			8		
		3						6
7			9			2		

SUDOKU 365

DATE: _____ TIME: _____

Question 309

			4					
		8		7		2		
3								
		6		3		4		
								3
		3		1		6		
							1	

SUD365OKU

Question 310

						5		
	6		8	4			7	
							2	
	2			6			1	
	7							
	1			7	4		5	
		6						

Question 311

6				4				9
		4				5		
9				6				4
		1				6		
5				2				7

Question 312

		6	4					9
4					1		2	
								8
						2		
	7				6			
		1			4		5	
6	2							
8	9		7		5	6		

DATE: _____ TIME: _____

Question 313

		4	9		2			
							9	
2		6				4	7	
4	5		6					
				4	6			5
1	8		2					
							2	
				8				6

DATE: TIME:

Question 314

		2				7		
							2	
5								
9								
				9				
								4
								2
	9							
		4				9		

SUD 365 OKU

DATE: _____ TIME: _____

Question 315

	3						2	
		5	8		1	7		
7				3				6
		9	4		7	3		
	2						3	

SUD●OKU 365

Question 316

								6
		5						
		1			9			
8								
2	6						9	1
					1			
		6						
		8						
			9					

Question **317**

7								2
			3	1				
				5			9	
	2		9		6		7	
	5			2				
				9	4			
9								1

SUD 365 OKU

Question 318

	4		9	8	5		3	
6	9	5				2	7	3
	5		2	3	6		4	

DATE: TIME:

Question 319

				6				
	2			7			6	
				1				
2	9	7				6	1	5
				4				
	5			8			7	
				2				

SUD 365 OKU

DATE: TIME:

Question 320

				9				
	2						9	
			2					
9		7				1		3
			3					
	3						1	
			1					

DATE: _____ TIME: _____

Question 321

	7			5			2	
		9				3		
	1			5			9	
		2				5		
	4			3			7	

Question 322

6								1
	7			9			2	
	6			4			3	
	8			5			4	
7								6

SUDOKU 365

Question 323

							7	
5	7			6			8	
	2			9			6	
	3			4			9	5
	4							

SUD 365 OKU

DATE: _____   TIME: _____

				4				
	5			1			8	
5	8			9			1	6
	2			7			3	
				2				

DATE: TIME:

Question 325

	7			4			3	
		3				2		
	9			1			6	
		5				7		
	6			5			8	

SUD365OKU

Question 326

				8				
8						4		6
		8						
6				4				7
						3		
9		5						2
				9				

345

Question 327

				3				
1								5
				5				
4		5				6		3
				9				
9								8
				2				

DATE: TIME:

Question 328

6				5		9		
							4	
						1	7	
	2			8	4			7
	1		2					
						2		
						7		1
1	9	2	8					4
							2	

DATE: _____ TIME: _____

Question **329**

								1
5						3		2
	6	2	1		4			
6								
3		8		6				
		9			7			6
		3	4				7	
						8	6	
		6	3					

DATE: TIME:

Question 330

7				6				5
	5			7			8	
3				1				9
	3			8			5	
8				2				6

SUD 365 OKU

Question 331

4								1
	8			3			6	
	1			8			9	
	4			7			1	

DATE: _____ TIME: _____

Question 332

	7			5			4	
	2			4			5	
	6			2			1	
8								2

DATE: TIME:

Question 333

				1				
	3			8			4	
	1			4			2	
	6			8			3	
				3				

Question **334**

	2			1			7	
7	8			3			6	5
	4			8			1	

SUD365OKU

DATE: TIME:

Question 335

1								
	4			7			2	
	6			4			7	
	8			9			5	
								9

Question 336

	6			7			3	
	1			9			7	
	9			6			1	
	5						9	

DATE: _____ TIME: _____

Question 337

	7						5	
	8			7			3	
	1			9			6	
	3			1			2	

SUD 365 OKU

DATE: TIME:

Question 338

				9				
3				5				2
4				2				6
8				4				3
				6				

9				7				6
				3				
3				5				4
				1				
2				8				5

Question 340

	1						3	
				1				
		3						
1				3				5
						9		
				9				
	7						2	

DATE: _____ TIME: _____

Question 341

	6			3			8	
				8				
	1						7	
				9				
	4			5			9	

DATE: _____ TIME: _____

Question 342

	2			1			6	
	3		2		1		9	
	6			8			3	

DATE: TIME:

Question 343

	4						9	
				6				
2			6		4			9
				2				
	1						8	

Question 344

				2				
1								8
				5				
7								5
				4				
2								6
				6				

DATE: TIME:

Question 345

	4			1			2	
	3			8			9	
		1						
	1			9			7	

DATE: _____ TIME: _____

Question 346

	8			1			3	
					6			
	3			7			6	
	5			3			8	

DATE: TIME:

Question 347

	2			1			8	
		9						
	7			2			3	
	5			9			7	

Question 348

	9			5			8	
	4			2			3	
				7				
	5			6			1	

DATE: _____ TIME: _____

Question **349**

	8			2			1	
				4				
	7			1			5	
	6			7			3	

Question 350

	5			7			9	
	6			5			1	
				1				
	3			6			4	

Question 351

	8			2			1	
	6			7			8	
	1			9			4	

DATE: TIME:

Question 352

	9			2			3	
	3			4			1	
	7			3			6	

Question 353

	4			5			7	
	1			6			9	
	3			9			2	

DATE: TIME:

Question 354

2				8				4
5				4				9
8				2				6

DATE: TIME:

Question 355

		2		7		6		
		7		5		2		
		4		1		9		

Question 356

	2			6			9	
	3			5			4	
	5			2			1	

Question 357

2				1				6
3				2				1
9				5				8

Question 358

	8				3			
		9		1				
			2					
		3		6				
	2				8			

DATE: TIME:

Question **359**

2								4
				9				
	2							
						8		
		2						
								6
			1					

DATE: TIME:

Question 360

	6					9		
2				9				7
		3					6	

DATE: _____   TIME: _____

Question 361

			4					
	4						3	
				9				
9								
						3	8	7
	1		6			9	5	2

DATE: TIME:

Question 362

9	4	5						
2	1	6					4	
				1			7	
	3			9			6	

DATE: TIME:

Question 363

	7						5	
	2		5	4	8		3	
	9						6	

Question **364**

	4						5	
			7					
	3						6	

Question 365

			3		9			
			2		4			

브레인 트레이닝 스도쿠 365

ANSWER

SUD정답OKU

8	9	1	5	4	6	7	2	3
5	3	2	1	7	9	8	6	4
7	4	6	2	8	3	1	5	9
9	5	8	7	6	4	2	3	1
1	2	4	3	9	5	6	7	8
6	7	3	8	2	1	4	9	5
3	8	7	9	1	2	5	4	6
4	1	9	6	5	7	3	8	2
2	6	5	4	3	8	9	1	7

1	3	9	8	6	7	5	2	4
8	4	6	2	5	3	9	7	1
7	2	5	4	9	1	8	3	6
6	9	2	7	3	5	4	1	8
3	1	8	6	2	4	7	5	9
5	7	4	1	8	9	3	6	2
2	6	7	3	4	8	1	9	5
4	5	3	9	1	6	2	8	7
9	8	1	5	7	2	6	4	3

1	9	3	2	8	7	5	6	4
7	6	8	4	9	5	1	3	2
2	5	4	1	3	6	7	9	8
3	1	7	5	6	8	4	2	9
9	8	2	3	1	4	6	7	5
6	4	5	7	2	9	3	8	1
4	3	9	6	5	2	8	1	7
8	7	6	9	4	1	2	5	3
5	2	1	8	7	3	9	4	6

6	4	5	9	2	3	8	7	1
7	8	2	6	1	5	9	3	4
3	9	1	7	8	4	5	2	6
5	1	8	3	4	6	2	9	7
4	3	9	2	7	8	1	6	5
2	7	6	1	5	9	3	4	8
8	6	4	5	9	2	7	1	3
9	5	7	4	3	1	6	8	2
1	2	3	8	6	7	4	5	9

1	9	3	6	4	7	2	8	5
2	8	4	9	1	5	3	6	7
5	6	7	2	3	8	1	9	4
6	1	5	3	8	4	9	7	2
3	4	9	7	2	6	8	5	1
7	2	8	5	9	1	4	3	6
4	3	6	8	7	2	5	1	9
9	7	1	4	5	3	6	2	8
8	5	2	1	6	9	7	4	3

7	5	2	3	4	6	1	9	8
8	4	1	7	9	2	3	6	5
6	9	3	8	5	1	2	7	4
9	6	4	2	1	7	8	5	3
2	3	5	6	8	9	4	1	7
1	7	8	4	3	5	6	2	9
5	8	9	1	6	3	7	4	2
4	2	6	5	7	8	9	3	1
3	1	7	9	2	4	5	8	6

SUD정답OKU

Answer 7

9	5	3	1	7	2	4	8	6
1	7	6	4	3	8	5	9	2
8	2	4	5	6	9	1	3	7
5	8	1	2	4	6	9	7	3
3	9	2	7	5	1	8	6	4
4	6	7	8	9	3	2	5	1
6	3	5	9	2	4	7	1	8
7	4	8	6	1	5	3	2	9
2	1	9	3	8	7	6	4	5

Answer 8

7	9	8	3	4	5	1	2	6
5	3	4	6	2	1	7	8	9
1	6	2	9	7	8	5	4	3
6	2	5	7	8	3	4	9	1
9	1	3	5	6	4	8	7	2
4	8	7	2	1	9	6	3	5
2	5	1	8	9	7	3	6	4
3	7	6	4	5	2	9	1	8
8	4	9	1	3	6	2	5	7

Answer 9

8	3	1	6	7	2	9	4	5
7	4	9	8	5	1	6	2	3
6	5	2	9	4	3	7	1	8
4	7	6	1	8	9	3	5	2
1	2	3	5	6	4	8	9	7
9	8	5	2	3	7	4	6	1
2	6	7	3	9	5	1	8	4
5	9	4	7	1	8	2	3	6
3	1	8	4	2	6	5	7	9

Answer 10

3	8	9	1	5	4	2	7	6
1	6	7	3	2	9	4	5	8
4	2	5	6	8	7	3	1	9
6	4	1	9	3	5	8	2	7
7	3	8	2	4	1	9	6	5
5	9	2	7	6	8	1	4	3
9	1	6	8	7	2	5	3	4
2	5	3	4	9	6	7	8	1
8	7	4	5	1	3	6	9	2

Answer 11

3	1	9	7	5	8	6	4	2
6	2	7	3	4	1	5	9	8
5	8	4	2	9	6	1	3	7
2	6	5	1	3	9	8	7	4
7	4	3	8	6	5	2	1	9
1	9	8	4	7	2	3	6	5
8	5	6	9	1	4	7	2	3
4	7	1	5	2	3	9	8	6
9	3	2	6	8	7	4	5	1

Answer 12

3	9	1	4	5	6	2	8	7
6	2	5	8	1	7	9	4	3
7	4	8	3	9	2	1	5	6
9	3	6	5	7	1	4	2	8
2	8	4	6	3	9	5	7	1
5	1	7	2	8	4	3	6	9
8	6	3	1	2	5	7	9	4
4	7	2	9	6	3	8	1	5
1	5	9	7	4	8	6	3	2

SUD 정답 OKU

Answer 13

7	2	1	9	5	4	3	6	8
8	4	9	6	1	3	7	2	5
3	6	5	2	7	8	9	4	1
2	8	6	5	3	9	4	1	7
1	3	7	4	8	6	5	9	2
5	9	4	1	2	7	6	8	3
9	5	8	7	6	2	1	3	4
4	1	3	8	9	5	2	7	6
6	7	2	3	4	1	8	5	9

Answer 14

5	1	9	2	8	4	3	6	7
8	2	7	6	9	3	1	5	4
6	4	3	5	1	7	2	9	8
4	3	6	7	2	8	9	1	5
9	8	2	3	5	1	4	7	6
1	7	5	9	4	6	8	3	2
7	9	8	4	3	5	6	2	1
2	6	1	8	7	9	5	4	3
3	5	4	1	6	2	7	8	9

Answer 15

8	9	7	6	4	2	3	1	5
3	1	6	5	8	7	4	9	2
4	2	5	3	9	1	7	6	8
1	7	2	4	3	8	6	5	9
9	8	3	1	6	5	2	7	4
6	5	4	2	7	9	8	3	1
7	3	8	9	1	4	5	2	6
5	4	9	7	2	6	1	8	3
2	6	1	8	5	3	9	4	7

Answer 16

8	1	2	3	4	9	6	5	7
5	4	7	8	1	6	3	2	9
6	3	9	5	2	7	4	8	1
9	8	4	7	6	2	1	3	5
1	7	6	4	5	3	8	9	2
2	5	3	1	9	8	7	6	4
3	9	1	6	7	5	2	4	8
7	6	5	2	8	4	9	1	3
4	2	8	9	3	1	5	7	6

Answer 17

8	4	7	2	5	3	6	1	9
1	2	5	9	6	4	3	8	7
3	9	6	1	7	8	2	4	5
7	1	2	8	3	5	4	9	6
6	3	9	7	4	1	8	5	2
4	5	8	6	9	2	1	7	3
5	7	4	3	1	6	9	2	8
9	8	3	4	2	7	5	6	1
2	6	1	5	8	9	7	3	4

Answer 18

1	4	7	3	9	5	6	2	8
5	9	8	2	7	6	3	1	4
6	3	2	8	4	1	5	9	7
8	6	5	7	1	3	9	4	2
3	1	4	9	8	2	7	5	6
7	2	9	6	5	4	1	8	3
2	5	3	1	6	8	4	7	9
4	7	6	5	2	9	8	3	1
9	8	1	4	3	7	2	6	5

SUD 정답 OKU

Answer 19

5	8	2	3	6	7	9	4	1
7	6	1	8	9	4	2	5	3
4	9	3	5	2	1	7	8	6
1	5	9	2	7	8	6	3	4
2	3	8	4	1	6	5	9	7
6	7	4	9	3	5	1	2	8
9	1	5	7	8	3	4	6	2
3	4	6	1	5	2	8	7	9
8	2	7	6	4	9	3	1	5

Answer 20

9	4	5	6	1	8	7	2	3
1	8	3	2	7	9	6	5	4
7	6	2	4	3	5	8	9	1
3	1	6	8	9	4	2	7	5
8	5	4	3	2	7	1	6	9
2	9	7	1	5	6	4	3	8
5	2	8	9	6	1	3	4	7
4	3	9	7	8	2	5	1	6
6	7	1	5	4	3	9	8	2

Answer 21

9	5	4	1	8	2	6	7	3
3	2	6	4	5	7	1	8	9
1	7	8	6	3	9	2	5	4
6	8	2	9	4	1	5	3	7
5	3	7	8	2	6	9	4	1
4	1	9	5	7	3	8	6	2
7	6	3	2	9	5	4	1	8
2	4	1	7	6	8	3	9	5
8	9	5	3	1	4	7	2	6

Answer 22

3	1	2	5	8	6	9	4	7
4	8	5	1	9	7	2	6	3
7	6	9	4	2	3	8	5	1
5	3	8	6	7	2	4	1	9
2	4	7	3	1	9	5	8	6
6	9	1	8	4	5	3	7	2
1	2	3	7	5	8	6	9	4
9	5	4	2	6	1	7	3	8
8	7	6	9	3	4	1	2	5

Answer 23

3	6	7	1	4	9	8	2	5
9	5	4	2	8	7	6	1	3
8	1	2	3	6	5	4	9	7
5	7	3	4	9	1	2	6	8
6	8	9	7	2	3	1	5	4
4	2	1	6	5	8	3	7	9
2	9	5	8	1	4	7	3	6
1	3	8	9	7	6	5	4	2
7	4	6	5	3	2	9	8	1

Answer 24

6	5	2	4	3	8	9	7	1
8	7	3	6	9	1	5	2	4
9	1	4	5	2	7	6	8	3
5	3	8	9	1	2	4	6	7
1	6	7	8	4	5	2	3	9
2	4	9	3	7	6	8	1	5
4	2	5	1	6	3	7	9	8
3	8	6	7	5	9	1	4	2
7	9	1	2	8	4	3	5	6

SUD정답OKU

Answer 25

4	9	2	7	8	3	1	6	5
7	5	1	6	9	2	4	3	8
6	8	3	1	4	5	2	9	7
9	2	7	4	3	1	5	8	6
1	4	6	8	5	9	7	2	3
8	3	5	2	7	6	9	1	4
3	7	4	9	1	8	6	5	2
2	1	8	5	6	7	3	4	9
5	6	9	3	2	4	8	7	1

Answer 26

2	8	3	5	4	6	9	7	1
5	4	1	9	7	3	8	2	6
6	7	9	8	1	2	3	4	5
1	2	8	7	5	9	4	6	3
3	9	6	4	2	1	7	5	8
7	5	4	3	6	8	1	9	2
4	3	2	1	9	5	6	8	7
9	1	5	6	8	7	2	3	4
8	6	7	2	3	4	5	1	9

Answer 27

8	2	7	1	3	9	6	5	4
6	4	3	7	5	2	8	9	1
5	9	1	8	6	4	2	7	3
4	3	5	2	8	6	9	1	7
2	7	6	9	4	1	3	8	5
1	8	9	3	7	5	4	2	6
9	6	2	5	1	3	7	4	8
7	5	4	6	9	8	1	3	2
3	1	8	4	2	7	5	6	9

Answer 28

5	6	3	9	2	7	1	4	8
7	4	8	1	6	3	2	5	9
1	9	2	5	8	4	7	6	3
6	1	9	7	3	5	4	8	2
8	3	5	2	4	1	6	9	7
2	7	4	6	9	8	5	3	1
4	2	7	3	5	9	8	1	6
3	5	6	8	1	2	9	7	4
9	8	1	4	7	6	3	2	5

Answer 29

4	6	2	5	7	1	8	3	9
1	9	5	3	2	8	7	4	6
7	3	8	6	9	4	5	2	1
3	5	9	4	1	6	2	7	8
8	4	1	2	3	7	9	6	5
2	7	6	9	8	5	3	1	4
5	2	4	7	6	9	1	8	3
9	1	3	8	4	2	6	5	7
6	8	7	1	5	3	4	9	2

Answer 30

7	4	3	9	6	5	8	2	1
2	8	6	1	4	3	9	5	7
9	5	1	7	8	2	6	4	3
4	1	8	2	3	6	5	7	9
3	2	9	5	7	8	4	1	6
5	6	7	4	9	1	2	3	8
8	3	2	6	5	7	1	9	4
6	9	5	3	1	4	7	8	2
1	7	4	8	2	9	3	6	5

SUD정답OKU

Answer 31

4	7	3	5	1	9	6	8	2
6	5	9	4	8	2	1	3	7
1	2	8	3	6	7	4	9	5
5	9	1	8	4	6	2	7	3
2	8	6	7	3	5	9	4	1
7	3	4	2	9	1	5	6	8
3	6	5	1	7	4	8	2	9
9	1	7	6	2	8	3	5	4
8	4	2	9	5	3	7	1	6

Answer 32

3	5	8	1	6	9	4	2	7
2	9	7	5	8	4	3	1	6
4	1	6	7	2	3	5	8	9
5	8	3	9	1	6	2	7	4
9	2	1	4	7	5	6	3	8
6	7	4	8	3	2	9	5	1
7	6	5	3	4	1	8	9	2
8	4	9	2	5	7	1	6	3
1	3	2	6	9	8	7	4	5

Answer 33

3	5	7	2	6	8	4	1	9
2	1	6	7	4	9	5	3	8
8	9	4	5	1	3	2	7	6
7	2	5	6	3	4	9	8	1
9	6	1	8	2	7	3	4	5
4	3	8	1	9	5	7	6	2
5	7	2	4	8	1	6	9	3
1	4	9	3	5	6	8	2	7
6	8	3	9	7	2	1	5	4

Answer 34

6	9	1	5	8	2	4	3	7
2	8	5	4	3	7	1	9	6
3	4	7	1	6	9	8	2	5
9	1	2	3	4	5	6	7	8
5	3	8	7	2	6	9	4	1
4	7	6	8	9	1	3	5	2
1	5	3	9	7	8	2	6	4
8	6	9	2	5	4	7	1	3
7	2	4	6	1	3	5	8	9

Answer 35

8	3	2	4	7	5	9	6	1
6	7	9	3	1	8	2	4	5
5	4	1	9	6	2	3	7	8
2	5	3	1	8	6	4	9	7
1	9	8	2	4	7	5	3	6
4	6	7	5	3	9	8	1	2
7	1	4	8	2	3	6	5	9
9	2	6	7	5	4	1	8	3
3	8	5	6	9	1	7	2	4

Answer 36

8	2	7	3	5	6	1	4	9
5	1	3	8	4	9	2	6	7
6	9	4	2	7	1	5	8	3
7	6	9	5	1	2	4	3	8
1	3	2	9	8	4	6	7	5
4	8	5	7	6	3	9	1	2
3	4	8	6	2	5	7	9	1
9	5	6	1	3	7	8	2	4
2	7	1	4	9	8	3	5	6

SUD정답OKU

Answer 37

2	4	7	3	5	6	8	1	9
5	1	6	8	2	9	4	7	3
9	8	3	4	1	7	6	2	5
6	2	9	1	7	8	5	3	4
3	7	8	5	6	4	1	9	2
1	5	4	9	3	2	7	8	6
7	3	5	6	9	1	2	4	8
4	6	2	7	8	3	9	5	1
8	9	1	2	4	5	3	6	7

Answer 38

7	8	9	3	2	6	4	1	5
6	4	1	9	5	7	8	2	3
2	3	5	4	1	8	6	7	9
5	1	6	7	9	4	3	8	2
4	2	7	8	3	1	5	9	6
3	9	8	5	6	2	1	4	7
1	5	3	2	4	9	7	6	8
9	7	4	6	8	5	2	3	1
8	6	2	1	7	3	9	5	4

Answer 39

5	2	9	4	6	8	1	3	7
7	3	4	9	2	1	5	6	8
1	6	8	3	7	5	2	9	4
3	1	7	2	8	4	9	5	6
9	8	6	5	3	7	4	2	1
2	4	5	1	9	6	7	8	3
8	5	3	7	1	9	6	4	2
6	9	1	8	4	2	3	7	5
4	7	2	6	5	3	8	1	9

Answer 40

7	2	5	4	6	1	8	9	3
9	6	3	5	8	2	4	1	7
4	8	1	3	7	9	6	5	2
2	1	9	8	3	4	7	6	5
8	3	7	6	9	5	2	4	1
6	5	4	2	1	7	3	8	9
5	7	8	9	2	6	1	3	4
3	9	2	1	4	8	5	7	6
1	4	6	7	5	3	9	2	8

Answer 41

8	4	6	5	9	2	3	1	7
5	3	1	7	6	4	8	2	9
7	9	2	1	3	8	4	6	5
1	8	5	3	2	9	7	4	6
3	6	7	4	1	5	2	9	8
9	2	4	8	7	6	1	5	3
6	5	8	2	4	3	9	7	1
2	7	9	6	8	1	5	3	4
4	1	3	9	5	7	6	8	2

Answer 42

2	9	6	5	1	7	8	4	3
1	4	7	9	3	8	6	5	2
5	3	8	4	2	6	9	1	7
3	2	4	6	8	9	5	7	1
8	5	9	1	7	3	4	2	6
7	6	1	2	4	5	3	9	8
6	8	2	7	9	4	1	3	5
9	1	3	8	5	2	7	6	4
4	7	5	3	6	1	2	8	9

SUD 정답 OKU

Answer 43

3	1	9	8	2	5	6	7	4
7	4	2	3	6	1	5	9	8
8	6	5	9	7	4	2	3	1
2	3	7	1	8	6	4	5	9
1	5	4	7	3	9	8	6	2
6	9	8	4	5	2	7	1	3
5	8	6	2	9	3	1	4	7
4	7	3	6	1	8	9	2	5
9	2	1	5	4	7	3	8	6

Answer 44

6	3	5	2	1	7	9	8	4
7	4	2	9	8	3	1	6	5
1	9	8	6	5	4	3	2	7
4	8	7	1	2	5	6	9	3
9	1	6	3	4	8	5	7	2
2	5	3	7	9	6	8	4	1
8	6	4	5	3	2	7	1	9
3	7	1	4	6	9	2	5	8
5	2	9	8	7	1	4	3	6

Answer 45

8	7	2	1	3	5	9	4	6
3	1	9	6	4	8	7	2	5
6	4	5	2	7	9	8	1	3
5	8	4	3	2	6	1	7	9
1	6	3	9	8	7	2	5	4
9	2	7	5	1	4	6	3	8
7	5	8	4	6	1	3	9	2
4	3	6	7	9	2	5	8	1
2	9	1	8	5	3	4	6	7

Answer 46

5	2	6	3	8	9	4	7	1
8	4	9	7	1	6	2	5	3
1	3	7	2	4	5	6	9	8
3	1	4	8	5	2	7	6	9
6	7	2	4	9	3	1	8	5
9	5	8	1	6	7	3	2	4
7	9	5	6	3	1	8	4	2
4	6	1	5	2	8	9	3	7
2	8	3	9	7	4	5	1	6

Answer 47

8	5	9	3	4	7	2	6	1
3	2	6	1	8	9	5	4	7
4	7	1	5	2	6	8	9	3
2	6	7	8	9	5	3	1	4
5	3	4	2	7	1	6	8	9
1	9	8	6	3	4	7	5	2
9	4	3	7	6	8	1	2	5
7	8	5	9	1	2	4	3	6
6	1	2	4	5	3	9	7	8

Answer 48

8	9	7	6	4	2	5	1	3
4	5	3	1	9	7	8	6	2
6	1	2	5	8	3	4	7	9
9	4	8	3	1	6	2	5	7
5	2	6	9	7	4	1	3	8
7	3	1	2	5	8	9	4	6
1	8	4	7	3	9	6	2	5
3	6	9	4	2	5	7	8	1
2	7	5	8	6	1	3	9	4

SUD정답OKU

3	1	2	6	4	5	8	9	7
4	8	9	1	7	3	2	5	6
5	7	6	2	8	9	4	1	3
1	9	3	4	2	6	5	7	8
2	6	7	5	9	8	1	3	4
8	5	4	7	3	1	9	6	2
6	3	1	8	5	4	7	2	9
7	4	5	9	6	2	3	8	1
9	2	8	3	1	7	6	4	5

9	5	3	2	4	7	6	1	8
6	8	4	5	3	1	2	9	7
7	1	2	8	9	6	3	5	4
1	2	7	4	6	5	8	3	9
3	4	8	9	1	2	7	6	5
5	9	6	3	7	8	1	4	2
8	3	9	6	2	4	5	7	1
2	6	1	7	5	9	4	8	3
4	7	5	1	8	3	9	2	6

8	5	9	7	4	6	1	2	3
7	3	2	1	5	8	4	9	6
4	1	6	9	2	3	5	8	7
6	9	8	4	3	1	7	5	2
5	2	3	8	7	9	6	4	1
1	7	4	5	6	2	8	3	9
9	6	5	2	1	4	3	7	8
3	8	7	6	9	5	2	1	4
2	4	1	3	8	7	9	6	5

1	4	3	9	6	8	5	2	7
5	7	9	2	3	4	8	1	6
6	8	2	1	5	7	9	4	3
8	3	7	6	2	1	4	9	5
2	9	5	8	4	3	7	6	1
4	1	6	7	9	5	3	8	2
7	2	8	3	1	9	6	5	4
9	5	1	4	7	6	2	3	8
3	6	4	5	8	2	1	7	9

4	2	3	7	5	1	9	8	6
7	1	6	4	8	9	2	5	3
8	5	9	6	2	3	1	4	7
1	8	5	2	6	7	4	3	9
3	7	2	1	9	4	8	6	5
9	6	4	5	3	8	7	2	1
2	4	1	3	7	5	6	9	8
5	9	7	8	4	6	3	1	2
6	3	8	9	1	2	5	7	4

6	7	4	5	8	2	1	3	9
9	8	5	3	4	1	6	7	2
1	2	3	9	6	7	5	8	4
2	1	9	6	7	5	3	4	8
5	4	8	1	9	3	2	6	7
3	6	7	8	2	4	9	1	5
4	5	1	7	3	9	8	2	6
8	9	2	4	1	6	7	5	3
7	3	6	2	5	8	4	9	1

SUD정답OKU

Answer 55

5	3	9	2	8	1	7	4	6
2	7	6	3	5	4	1	9	8
4	8	1	9	6	7	2	5	3
6	9	4	8	1	2	5	3	7
3	1	7	6	4	5	8	2	9
8	5	2	7	9	3	4	6	1
1	6	8	5	2	9	3	7	4
7	4	5	1	3	6	9	8	2
9	2	3	4	7	8	6	1	5

Answer 56

1	5	3	6	8	7	4	9	2
9	8	6	4	3	2	7	5	1
2	4	7	9	1	5	3	8	6
3	9	4	2	5	6	1	7	8
7	2	5	8	4	1	9	6	3
6	1	8	3	7	9	2	4	5
5	6	9	1	2	4	8	3	7
4	3	2	7	6	8	5	1	9
8	7	1	5	9	3	6	2	4

Answer 57

9	2	5	8	6	7	1	4	3
3	8	4	2	1	9	5	7	6
1	6	7	5	4	3	8	2	9
5	4	2	9	3	6	7	1	8
8	7	9	1	2	5	3	6	4
6	1	3	7	8	4	2	9	5
4	3	8	6	7	1	9	5	2
7	9	6	3	5	2	4	8	1
2	5	1	4	9	8	6	3	7

Answer 58

4	6	8	7	5	9	3	1	2
7	2	5	8	1	3	6	4	9
3	9	1	4	6	2	5	7	8
9	1	3	2	8	4	7	6	5
2	5	4	3	7	6	9	8	1
8	7	6	1	9	5	4	2	3
1	8	9	6	3	7	2	5	4
5	4	7	9	2	8	1	3	6
6	3	2	5	4	1	8	9	7

Answer 59

4	6	2	5	1	7	8	3	9
8	3	9	4	6	2	5	1	7
1	7	5	3	9	8	6	2	4
2	4	7	9	5	1	3	6	8
3	9	8	2	7	6	1	4	5
5	1	6	8	3	4	9	7	2
7	8	3	1	2	9	4	5	6
6	5	4	7	8	3	2	9	1
9	2	1	6	4	5	7	8	3

Answer 60

4	2	6	1	8	5	9	3	7
9	5	3	2	7	6	1	8	4
8	1	7	4	9	3	6	2	5
7	9	1	5	4	8	3	6	2
5	3	4	6	2	9	7	1	8
2	6	8	3	1	7	5	4	9
6	4	2	9	5	1	8	7	3
3	8	9	7	6	4	2	5	1
1	7	5	8	3	2	4	9	6

SUD정답OKU

Answer 61

7	9	3	6	2	1	8	5	4
6	8	4	5	9	3	2	1	7
2	5	1	8	4	7	9	3	6
9	3	5	1	8	4	6	7	2
1	2	8	3	7	6	5	4	9
4	6	7	9	5	2	1	8	3
5	4	6	7	1	9	3	2	8
3	1	2	4	6	8	7	9	5
8	7	9	2	3	5	4	6	1

Answer 62

8	4	3	5	1	6	9	7	2
5	2	6	3	9	7	4	1	8
7	9	1	8	4	2	6	5	3
4	5	2	6	8	9	1	3	7
6	1	7	2	5	3	8	4	9
3	8	9	1	7	4	5	2	6
1	6	5	7	3	8	2	9	4
2	7	4	9	6	1	3	8	5
9	3	8	4	2	5	7	6	1

Answer 63

5	3	6	8	7	2	1	4	9
1	9	4	3	6	5	8	2	7
7	8	2	4	1	9	5	6	3
3	7	5	2	4	1	6	9	8
6	4	8	5	9	3	2	7	1
9	2	1	7	8	6	4	3	5
8	5	9	6	2	7	3	1	4
2	1	3	9	5	4	7	8	6
4	6	7	1	3	8	9	5	2

Answer 64

7	3	6	8	5	2	4	1	9
9	5	2	7	1	4	8	3	6
1	4	8	3	9	6	2	5	7
3	8	9	2	7	5	6	4	1
6	2	1	4	3	9	7	8	5
4	7	5	1	6	8	9	2	3
5	9	4	6	2	1	3	7	8
8	6	3	5	4	7	1	9	2
2	1	7	9	8	3	5	6	4

Answer 65

1	8	7	5	2	4	6	3	9
3	4	2	9	6	8	7	1	5
5	6	9	7	3	1	4	8	2
6	9	3	1	7	2	8	5	4
8	7	1	6	4	5	2	9	3
4	2	5	8	9	3	1	6	7
7	5	8	4	1	9	3	2	6
9	3	6	2	8	7	5	4	1
2	1	4	3	5	6	9	7	8

Answer 66

3	9	2	7	8	6	1	5	4
4	5	7	2	3	1	8	6	9
8	1	6	9	4	5	3	2	7
9	6	8	4	2	3	7	1	5
2	3	1	5	7	9	6	4	8
5	7	4	6	1	8	2	9	3
6	8	3	1	9	4	5	7	2
1	2	9	3	5	7	4	8	6
7	4	5	8	6	2	9	3	1

SUD 정답 OKU

Answer 67

4	2	6	8	5	9	7	3	1
7	9	8	2	1	3	6	5	4
5	3	1	7	6	4	2	9	8
1	7	5	3	9	6	8	4	2
3	4	2	1	8	7	5	6	9
6	8	9	4	2	5	1	7	3
2	6	4	9	7	1	3	8	5
9	1	7	5	3	8	4	2	6
8	5	3	6	4	2	9	1	7

Answer 68

6	3	5	9	1	8	2	7	4
2	1	8	5	4	7	6	9	3
7	9	4	2	3	6	1	5	8
8	5	9	4	7	2	3	6	1
4	2	3	6	5	1	7	8	9
1	7	6	8	9	3	5	4	2
3	6	1	7	8	9	4	2	5
9	4	2	1	6	5	8	3	7
5	8	7	3	2	4	9	1	6

Answer 69

1	6	9	8	7	5	4	3	2
5	7	8	4	2	3	9	6	1
3	4	2	6	1	9	7	5	8
4	8	3	9	6	1	5	2	7
6	9	5	2	8	7	3	1	4
2	1	7	3	5	4	6	8	9
7	5	4	1	3	8	2	9	6
9	2	1	5	4	6	8	7	3
8	3	6	7	9	2	1	4	5

Answer 70

5	2	8	7	3	1	6	4	9
9	4	7	8	2	6	5	3	1
3	6	1	9	5	4	2	8	7
4	8	6	1	9	3	7	5	2
1	5	9	2	4	7	3	6	8
7	3	2	5	6	8	1	9	4
6	7	5	4	1	9	8	2	3
8	9	3	6	7	2	4	1	5
2	1	4	3	8	5	9	7	6

Answer 71

2	8	6	3	1	5	7	4	9
7	1	5	9	4	2	6	3	8
4	9	3	7	8	6	5	1	2
9	4	1	6	7	8	2	5	3
8	6	7	2	5	3	4	9	1
3	5	2	1	9	4	8	6	7
6	7	4	8	3	1	9	2	5
5	3	9	4	2	7	1	8	6
1	2	8	5	6	9	3	7	4

Answer 72

1	9	2	8	6	5	3	7	4
8	6	4	7	1	3	5	9	2
5	7	3	2	4	9	1	8	6
7	5	8	6	3	2	4	1	9
2	4	9	1	7	8	6	3	5
3	1	6	5	9	4	7	2	8
4	8	1	9	5	7	2	6	3
9	3	7	4	2	6	8	5	1
6	2	5	3	8	1	9	4	7

SUDOKU

Answer 73

2	3	9	1	4	8	6	5	7
6	8	7	3	2	5	1	9	4
5	1	4	6	7	9	2	8	3
8	4	6	7	9	2	5	3	1
1	2	5	8	3	4	7	6	9
9	7	3	5	1	6	8	4	2
7	9	8	4	6	1	3	2	5
4	5	1	2	8	3	9	7	6
3	6	2	9	5	7	4	1	8

Answer 74

5	1	2	8	4	3	7	9	6
7	9	4	5	2	6	8	1	3
6	3	8	1	9	7	5	2	4
2	7	9	3	6	8	1	4	5
4	6	1	9	7	5	3	8	2
8	5	3	4	1	2	6	7	9
9	2	7	6	5	1	4	3	8
1	8	5	2	3	4	9	6	7
3	4	6	7	8	9	2	5	1

Answer 75

2	6	9	5	3	4	8	1	7
5	8	1	6	9	7	3	2	4
3	7	4	1	2	8	9	5	6
7	1	8	9	5	3	4	6	2
6	4	3	8	7	2	5	9	1
9	2	5	4	6	1	7	3	8
1	5	7	2	8	9	6	4	3
8	9	2	3	4	6	1	7	5
4	3	6	7	1	5	2	8	9

Answer 76

3	7	8	6	2	1	5	9	4
6	5	2	4	3	9	7	1	8
1	4	9	8	7	5	3	6	2
7	6	3	9	8	2	1	4	5
8	9	1	5	4	3	6	2	7
4	2	5	7	1	6	8	3	9
5	3	6	2	9	8	4	7	1
9	8	4	1	6	7	2	5	3
2	1	7	3	5	4	9	8	6

Answer 77

1	3	6	5	2	9	8	4	7
2	5	8	4	6	7	3	9	1
7	9	4	3	1	8	5	2	6
3	8	2	1	9	4	6	7	5
5	7	9	6	3	2	4	1	8
4	6	1	8	7	5	9	3	2
6	2	3	9	8	1	7	5	4
9	4	7	2	5	6	1	8	3
8	1	5	7	4	3	2	6	9

Answer 78

7	4	9	3	6	5	2	1	8
2	1	8	7	9	4	6	5	3
5	3	6	1	2	8	9	4	7
8	5	7	4	3	9	1	2	6
1	9	2	8	7	6	4	3	5
4	6	3	2	5	1	7	8	9
3	2	5	9	4	7	8	6	1
9	8	4	6	1	3	5	7	2
6	7	1	5	8	2	3	9	4

SUD 정답 OKU

Answer 79

5	4	2	3	8	1	9	6	7
9	8	7	2	6	4	5	3	1
6	3	1	5	9	7	4	2	8
1	9	5	8	2	6	3	7	4
3	7	8	4	1	9	6	5	2
2	6	4	7	5	3	1	8	9
7	1	3	6	4	8	2	9	5
8	5	9	1	3	2	7	4	6
4	2	6	9	7	5	8	1	3

Answer 80

5	6	3	9	2	7	1	4	8
7	4	8	1	6	3	2	5	9
1	9	2	5	8	4	7	6	3
6	1	9	7	3	5	4	8	2
8	3	5	2	4	1	6	9	7
2	7	4	6	9	8	5	3	1
4	2	7	3	5	9	8	1	6
3	5	6	8	1	2	9	7	4
9	8	1	4	7	6	3	2	5

Answer 81

4	7	1	8	2	3	6	5	9
8	2	9	5	1	6	4	7	3
5	3	6	7	9	4	2	8	1
3	9	5	6	7	8	1	4	2
7	8	4	2	3	1	9	6	5
1	6	2	4	5	9	8	3	7
2	5	8	1	6	7	3	9	4
9	4	7	3	8	2	5	1	6
6	1	3	9	4	5	7	2	8

Answer 82

4	9	6	7	5	8	3	2	1
1	8	3	6	4	2	5	9	7
5	7	2	3	1	9	8	4	6
3	4	9	5	2	1	7	6	8
6	1	7	8	3	4	2	5	9
2	5	8	9	6	7	4	1	3
9	3	1	4	7	5	6	8	2
7	2	5	1	8	6	9	3	4
8	6	4	2	9	3	1	7	5

Answer 83

6	7	1	9	3	8	4	5	2
2	8	5	4	6	7	9	1	3
3	9	4	2	1	5	8	6	7
5	4	6	7	8	2	3	9	1
9	1	8	6	5	3	2	7	4
7	3	2	1	4	9	5	8	6
8	6	7	3	9	4	1	2	5
1	5	3	8	2	6	7	4	9
4	2	9	5	7	1	6	3	8

Answer 84

6	8	9	1	3	4	7	2	5
7	3	4	5	2	6	8	9	1
1	2	5	9	8	7	3	4	6
9	5	2	4	1	3	6	7	8
4	7	1	6	9	8	5	3	2
3	6	8	7	5	2	9	1	4
8	4	7	3	6	1	2	5	9
5	1	6	2	7	9	4	8	3
2	9	3	8	4	5	1	6	7

SUD 정답 OKU

4	8	3	6	5	9	7	1	2
6	9	2	3	7	1	5	4	8
5	7	1	8	2	4	9	3	6
2	3	5	7	9	8	1	6	4
8	1	9	2	4	6	3	7	5
7	6	4	5	1	3	2	8	9
3	2	7	4	6	5	8	9	1
9	4	8	1	3	2	6	5	7
1	5	6	9	8	7	4	2	3

3	4	7	9	1	5	8	6	2
2	9	5	6	8	7	1	3	4
6	1	8	2	3	4	7	9	5
5	2	1	4	9	6	3	8	7
9	8	3	1	7	2	4	5	6
4	7	6	8	5	3	9	2	1
8	3	2	5	4	1	6	7	9
7	5	4	3	6	9	2	1	8
1	6	9	7	2	8	5	4	3

3	4	9	2	1	8	7	6	5
5	2	7	9	6	3	4	1	8
6	8	1	7	5	4	9	3	2
1	3	8	6	4	7	2	5	9
9	6	4	8	2	5	3	7	1
2	7	5	3	9	1	6	8	4
7	5	3	4	8	2	1	9	6
4	1	6	5	7	9	8	2	3
8	9	2	1	3	6	5	4	7

9	3	7	8	2	4	6	1	5
2	6	8	9	5	1	7	3	4
1	5	4	3	6	7	2	8	9
5	1	9	4	8	2	3	7	6
8	7	2	5	3	6	9	4	1
3	4	6	1	7	9	8	5	2
7	9	5	6	1	8	4	2	3
6	8	1	2	4	3	5	9	7
4	2	3	7	9	5	1	6	8

3	6	4	7	2	1	8	9	5
1	5	7	9	4	8	2	3	6
2	9	8	6	5	3	7	4	1
9	7	5	3	1	6	4	2	8
8	3	1	4	9	2	5	6	7
4	2	6	5	8	7	9	1	3
7	4	3	2	6	5	1	8	9
5	1	9	8	3	4	6	7	2
6	8	2	1	7	9	3	5	4

8	5	4	6	3	1	7	2	9
3	9	2	4	8	7	5	6	1
7	6	1	9	2	5	3	4	8
6	3	8	2	1	9	4	7	5
1	7	5	8	6	4	9	3	2
4	2	9	7	5	3	1	8	6
2	4	7	1	9	8	6	5	3
5	1	6	3	4	2	8	9	7
9	8	3	5	7	6	2	1	4

SUDOKU 정답

Answer 91

9	3	2	1	8	7	6	5	4
8	5	6	4	3	2	1	7	9
1	7	4	6	5	9	2	3	8
2	6	5	8	7	1	4	9	3
7	9	3	5	4	6	8	1	2
4	1	8	2	9	3	7	6	5
5	4	1	9	6	8	3	2	7
3	2	9	7	1	4	5	8	6
6	8	7	3	2	5	9	4	1

Answer 92

6	9	4	7	2	3	1	8	5
7	3	8	1	9	5	2	6	4
5	1	2	6	4	8	7	9	3
4	6	3	2	5	1	9	7	8
2	7	5	9	8	4	3	1	6
1	8	9	3	7	6	4	5	2
3	2	6	8	1	7	5	4	9
8	4	7	5	3	9	6	2	1
9	5	1	4	6	2	8	3	7

Answer 93

9	1	2	8	6	5	4	3	7
5	8	4	7	2	3	9	1	6
7	3	6	1	4	9	8	5	2
3	7	5	6	9	8	1	2	4
6	4	1	2	3	7	5	8	9
2	9	8	5	1	4	6	7	3
8	6	9	3	5	2	7	4	1
4	5	3	9	7	1	2	6	8
1	2	7	4	8	6	3	9	5

Answer 94

1	3	2	7	6	5	4	9	8
8	5	6	2	9	4	1	3	7
4	7	9	8	1	3	5	2	6
5	2	3	4	8	7	9	6	1
7	1	8	6	5	9	3	4	2
9	6	4	3	2	1	8	7	5
6	8	1	9	4	2	7	5	3
3	9	5	1	7	6	2	8	4
2	4	7	5	3	8	6	1	9

Answer 95

4	8	1	2	3	5	6	7	9
7	3	9	6	8	4	1	2	5
2	5	6	1	7	9	3	4	8
8	9	2	7	4	3	5	6	1
6	1	7	5	2	8	9	3	4
3	4	5	9	6	1	2	8	7
9	6	8	3	1	7	4	5	2
5	2	4	8	9	6	7	1	3
1	7	3	4	5	2	8	9	6

Answer 96

2	8	4	9	1	5	6	3	7
7	9	5	4	6	3	8	2	1
3	6	1	7	8	2	5	4	9
8	1	9	6	3	7	4	5	2
5	7	6	8	2	4	9	1	3
4	2	3	1	5	9	7	6	8
9	3	8	5	4	1	2	7	6
1	4	7	2	9	6	3	8	5
6	5	2	3	7	8	1	9	4

SUD정답OKU

Answer 97

3	7	5	6	8	9	1	4	2
6	1	9	5	4	2	7	3	8
4	8	2	7	1	3	5	9	6
5	3	1	2	6	7	9	8	4
2	4	6	8	9	5	3	1	7
7	9	8	1	3	4	6	2	5
1	6	3	4	5	8	2	7	9
9	2	4	3	7	6	8	5	1
8	5	7	9	2	1	4	6	3

Answer 98

3	5	7	9	6	1	4	2	8
2	8	1	5	4	3	7	6	9
6	9	4	2	7	8	1	3	5
9	1	6	7	3	2	8	5	4
7	4	3	8	5	6	9	1	2
5	2	8	4	1	9	3	7	6
4	6	5	1	8	7	2	9	3
1	3	2	6	9	4	5	8	7
8	7	9	3	2	5	6	4	1

Answer 99

1	5	4	6	2	8	3	9	7
9	8	2	5	3	7	4	1	6
7	3	6	1	4	9	8	2	5
5	4	7	3	6	2	1	8	9
6	9	1	7	8	4	2	5	3
8	2	3	9	5	1	6	7	4
4	7	9	8	1	3	5	6	2
3	1	5	2	7	6	9	4	8
2	6	8	4	9	5	7	3	1

Answer 100

7	2	1	4	3	6	9	5	8
4	8	6	5	9	2	1	7	3
9	5	3	8	7	1	2	4	6
6	4	7	2	8	3	5	1	9
3	1	2	9	5	7	8	6	4
5	9	8	6	1	4	7	3	2
1	6	9	7	4	8	3	2	5
2	7	5	3	6	9	4	8	1
8	3	4	1	2	5	6	9	7

Answer 101

3	7	6	2	5	1	9	4	8
8	2	4	9	3	6	5	7	1
5	9	1	7	8	4	6	3	2
7	1	5	8	4	9	3	2	6
9	6	2	1	7	3	8	5	4
4	8	3	6	2	5	7	1	9
1	5	8	4	6	7	2	9	3
6	4	7	3	9	2	1	8	5
2	3	9	5	1	8	4	6	7

Answer 102

7	5	9	2	8	1	6	4	3
1	6	3	9	4	5	2	7	8
8	2	4	3	6	7	1	9	5
9	3	8	1	7	6	4	5	2
2	4	1	8	5	9	3	6	7
5	7	6	4	3	2	8	1	9
3	8	5	7	1	4	9	2	6
6	1	2	5	9	3	7	8	4
4	9	7	6	2	8	5	3	1

SUD 정답 OKU

Answer 103

6	1	4	5	2	7	3	9	8
5	2	3	9	4	8	7	6	1
9	8	7	3	1	6	2	4	5
7	9	2	4	8	3	5	1	6
3	4	1	2	6	5	8	7	9
8	5	6	1	7	9	4	3	2
4	7	8	6	9	2	1	5	3
1	6	5	8	3	4	9	2	7
2	3	9	7	5	1	6	8	4

Answer 104

8	1	3	7	5	2	9	6	4
5	7	4	3	9	6	8	1	2
9	6	2	4	1	8	5	7	3
4	9	5	6	8	1	3	2	7
6	3	8	2	7	4	1	9	5
1	2	7	5	3	9	6	4	8
2	4	1	8	6	5	7	3	9
3	5	9	1	4	7	2	8	6
7	8	6	9	2	3	4	5	1

Answer 105

4	3	9	5	8	6	2	7	1
2	1	6	4	3	7	8	9	5
5	7	8	9	1	2	6	4	3
1	6	4	8	5	9	7	3	2
8	9	3	2	7	1	4	5	6
7	2	5	3	6	4	9	1	8
3	5	7	6	9	8	1	2	4
6	4	1	7	2	3	5	8	9
9	8	2	1	4	5	3	6	7

Answer 106

9	2	3	5	4	7	6	8	1
6	1	4	8	9	3	2	5	7
8	5	7	1	6	2	9	3	4
7	9	5	2	8	4	3	1	6
3	4	2	6	7	1	8	9	5
1	8	6	3	5	9	7	4	2
5	3	9	4	2	6	1	7	8
2	7	8	9	1	5	4	6	3
4	6	1	7	3	8	5	2	9

Answer 107

6	4	8	7	3	9	1	2	5
9	2	3	1	8	5	4	7	6
7	5	1	4	6	2	3	8	9
2	8	7	3	4	6	9	5	1
4	9	5	2	1	7	6	3	8
3	1	6	5	9	8	7	4	2
8	3	9	6	5	4	2	1	7
5	7	4	9	2	1	8	6	3
1	6	2	8	7	3	5	9	4

Answer 108

9	8	5	6	2	7	3	4	1
4	2	3	9	1	5	6	7	8
7	6	1	4	8	3	5	9	2
5	9	4	3	7	2	8	1	6
1	7	6	8	5	4	2	3	9
8	3	2	1	9	6	7	5	4
3	4	7	2	6	1	9	8	5
6	1	8	5	3	9	4	2	7
2	5	9	7	4	8	1	6	3

SUD 정답 OKU

Answer 109

2	3	5	1	7	8	6	9	4
4	1	8	5	6	9	3	2	7
9	6	7	4	3	2	8	5	1
3	7	1	6	9	4	5	8	2
8	4	6	3	2	5	1	7	9
5	2	9	7	8	1	4	6	3
1	9	3	8	5	7	2	4	6
7	5	4	2	1	6	9	3	8
6	8	2	9	4	3	7	1	5

Answer 110

1	3	9	7	8	4	5	2	6
7	2	5	6	9	3	8	1	4
4	8	6	1	5	2	3	7	9
6	4	2	8	1	9	7	5	3
5	9	7	2	3	6	1	4	8
8	1	3	4	7	5	6	9	2
2	6	1	5	4	8	9	3	7
9	7	8	3	2	1	4	6	5
3	5	4	9	6	7	2	8	1

Answer 111

1	7	4	8	6	5	9	3	2
5	3	2	7	1	9	4	6	8
9	8	6	4	3	2	1	5	7
6	5	9	2	4	1	8	7	3
8	1	3	6	5	7	2	9	4
4	2	7	3	9	8	5	1	6
7	4	1	9	2	6	3	8	5
3	9	8	5	7	4	6	2	1
2	6	5	1	8	3	7	4	9

Answer 112

2	3	9	5	7	1	6	8	4
6	8	1	9	2	4	3	7	5
4	5	7	3	6	8	1	2	9
8	1	3	4	5	6	7	9	2
7	9	6	2	8	3	5	4	1
5	2	4	1	9	7	8	3	6
9	6	2	8	3	2	4	1	7
3	4	5	7	1	5	9	6	8
1	7	8	6	4	9	2	5	3

Answer 113

4	3	9	5	8	6	2	7	1
2	1	6	4	3	7	8	9	5
5	7	8	9	1	2	6	4	3
1	6	4	8	5	9	7	3	2
8	9	3	2	7	1	4	5	6
7	2	5	3	6	4	9	1	8
3	5	7	6	9	8	1	2	4
6	4	1	7	2	3	5	8	9
9	8	2	1	4	5	3	6	7

Answer 114

4	5	7	9	8	6	3	2	1
1	2	6	7	5	3	4	9	8
9	3	8	2	4	1	5	6	7
2	1	9	5	6	7	8	4	3
3	6	5	4	1	8	2	7	9
8	7	4	3	2	9	1	5	6
5	8	1	6	7	2	9	3	4
6	9	2	1	3	4	7	8	5
7	4	3	8	9	5	6	1	2

SUD정답OKU

Answer 115

6	7	8	1	2	9	4	3	5
4	9	5	6	3	8	2	7	1
3	1	2	4	5	7	6	8	9
8	6	4	5	7	1	3	9	2
1	3	7	9	8	2	5	4	6
2	5	9	3	4	6	8	1	7
5	2	3	7	9	4	1	6	8
7	4	1	8	6	5	9	2	3
9	8	6	2	1	3	7	5	4

Answer 116

8	1	3	2	6	9	5	7	4
2	5	9	3	4	7	6	8	1
7	6	4	8	1	5	9	2	3
9	4	7	6	8	3	1	5	2
3	8	1	4	5	2	7	9	6
5	2	6	9	7	1	4	3	8
1	3	2	7	9	6	8	4	5
6	7	8	5	3	4	2	1	9
4	9	5	1	2	8	3	6	7

Answer 117

2	5	3	1	9	8	6	7	4
6	9	1	7	4	2	8	5	3
4	7	8	3	5	6	1	2	9
8	1	4	2	3	5	9	6	7
5	6	7	4	1	9	3	8	2
9	3	2	6	8	7	4	1	5
1	4	6	5	2	3	7	9	8
7	8	5	9	6	4	2	3	1
3	2	9	8	7	1	5	4	6

Answer 118

5	3	7	9	1	4	6	8	2
1	9	6	8	2	3	7	4	5
4	8	2	5	6	7	9	1	3
7	1	4	3	8	2	5	6	9
2	5	8	1	9	6	4	3	7
9	6	3	7	4	5	1	2	8
8	4	9	2	5	1	3	7	6
3	2	1	6	7	9	8	5	4
6	7	5	4	3	8	2	9	1

Answer 119

2	6	9	3	8	4	1	7	5
3	5	7	1	2	9	4	6	8
4	8	1	5	6	7	3	9	2
9	3	5	8	4	1	6	2	7
8	1	2	6	7	3	9	5	4
6	7	4	2	9	5	8	1	3
1	2	6	7	3	8	5	4	9
5	9	3	4	1	2	7	8	6
7	4	8	9	5	6	2	3	1

Answer 120

4	9	3	2	6	7	5	8	1
1	8	2	9	4	5	6	3	7
6	5	7	3	1	8	2	9	4
8	1	9	5	3	2	4	7	6
3	2	6	1	7	4	9	5	8
7	4	5	6	8	9	1	2	3
2	7	8	4	5	1	3	6	9
5	3	1	8	9	6	7	4	2
9	6	4	7	2	3	8	1	5

SUD 정답 OKU

Answer 121

5	8	7	1	6	3	9	4	2
9	6	1	8	4	2	7	5	3
4	3	2	5	7	9	1	8	6
8	4	6	7	3	5	2	9	1
1	2	5	9	8	6	4	3	7
3	7	9	2	1	4	8	6	5
7	1	3	4	5	8	6	2	9
6	9	8	3	2	7	5	1	4
2	5	4	6	9	1	3	7	8

Answer 122

4	3	1	5	9	6	2	7	8
8	6	5	1	2	7	3	4	9
7	2	9	3	4	8	1	6	5
1	9	3	6	5	4	7	8	2
6	7	2	8	1	3	5	9	4
5	8	4	2	7	9	6	3	1
9	1	8	7	6	5	4	2	3
3	5	7	4	8	2	9	1	6
2	4	6	9	3	1	8	5	7

Answer 123

9	5	1	7	3	6	2	8	4
3	2	8	4	9	5	7	1	6
7	4	6	8	2	1	9	5	3
6	8	4	1	5	9	3	2	7
1	3	9	2	4	7	5	6	8
2	7	5	3	6	8	4	9	1
8	9	2	6	7	3	1	4	5
4	6	3	5	1	2	8	7	9
5	1	7	9	8	4	6	3	2

Answer 124

8	3	5	1	7	4	9	6	2
2	4	6	8	5	9	3	7	1
1	9	7	2	3	6	8	4	5
7	5	3	6	2	1	4	8	9
9	6	1	4	8	7	2	5	3
4	2	8	3	9	5	7	1	6
6	8	9	5	4	3	1	2	7
3	1	2	7	6	8	5	9	4
5	7	4	9	1	2	6	3	8

Answer 125

1	4	2	8	7	5	6	9	3
9	7	5	3	2	6	8	4	1
8	3	6	4	1	9	2	5	7
5	2	8	1	4	7	3	6	9
7	6	3	5	9	2	1	8	4
4	1	9	6	8	3	5	7	2
3	9	4	2	6	8	7	1	5
2	8	1	7	5	4	9	3	6
6	5	7	9	3	1	4	2	8

Answer 126

2	7	6	4	1	8	3	9	5
8	3	5	7	9	2	1	6	4
9	1	4	6	5	3	2	7	8
3	4	7	2	6	1	8	5	9
1	5	8	3	7	9	4	2	6
6	9	2	8	4	5	7	1	3
4	2	1	5	8	6	9	3	7
7	6	3	9	2	4	5	8	1
5	8	9	1	3	7	6	4	2

SUD정답OKU

9	1	4	6	7	5	2	3	8
3	5	6	8	9	2	4	1	7
8	7	2	1	3	4	6	9	5
7	2	5	3	6	1	8	4	9
4	8	1	2	5	9	3	7	6
6	3	9	7	4	8	1	5	2
5	4	3	9	8	6	7	2	1
2	6	7	5	1	3	9	8	4
1	9	8	4	2	7	5	6	3

5	8	2	6	9	7	4	3	1
3	6	7	2	1	4	5	8	9
9	4	1	8	5	3	6	7	2
1	5	8	3	6	2	7	9	4
4	3	6	5	7	9	2	1	8
2	7	9	4	8	1	3	6	5
7	9	3	1	4	5	8	2	6
6	1	4	7	2	8	9	5	3
8	2	5	9	3	6	1	4	7

4	1	6	8	7	3	9	2	5
3	9	5	1	6	2	7	8	4
7	8	2	4	5	9	6	3	1
8	5	4	7	1	6	2	9	3
9	6	3	5	2	4	1	7	8
2	7	1	3	9	8	5	4	6
6	3	9	2	8	5	4	1	7
1	2	8	6	4	7	3	5	9
5	4	7	9	3	1	8	6	2

1	7	9	3	4	6	2	8	5
3	4	8	2	9	5	6	1	7
5	6	2	7	8	1	9	4	3
7	5	6	8	1	9	4	3	2
8	3	4	5	2	7	1	6	9
2	9	1	6	3	4	5	7	8
6	2	5	1	7	3	8	9	4
4	8	3	9	6	2	7	5	1
9	1	7	4	5	8	3	2	6

6	9	8	1	7	2	3	4	5
5	1	2	9	3	4	8	7	6
3	4	7	8	5	6	2	9	1
7	5	9	6	2	3	4	1	8
1	6	4	5	8	9	7	2	3
8	2	3	4	1	7	6	5	9
9	7	1	2	6	8	5	3	4
2	8	5	3	4	1	9	6	7
4	3	6	7	9	5	1	8	2

8	9	5	1	4	2	7	3	6
7	2	3	8	6	9	5	1	4
1	6	4	3	7	5	2	8	9
2	7	6	4	9	8	3	5	1
4	1	8	5	2	3	9	6	7
5	3	9	6	1	7	8	4	2
6	8	2	9	5	1	4	7	3
9	5	1	7	3	4	6	2	8
3	4	7	2	8	6	1	9	5

SUD정답OKU

Answer 133

2	9	1	7	5	4	6	8	3
8	4	6	2	9	3	1	5	7
5	3	7	1	6	8	4	2	9
3	7	4	6	8	5	9	1	2
1	2	5	9	4	7	3	6	8
6	8	9	3	1	2	7	4	5
7	6	3	5	2	1	8	9	4
9	5	8	4	7	6	2	3	1
4	1	2	8	3	9	5	7	6

Answer 134

6	4	9	8	1	7	2	5	3
7	5	8	9	2	3	6	4	1
1	2	3	6	5	4	7	8	9
4	3	2	7	9	8	1	6	5
5	6	7	1	4	2	3	9	8
8	9	1	3	6	5	4	2	7
9	1	4	5	3	6	8	7	2
2	8	5	4	7	1	9	3	6
3	7	6	2	8	9	5	1	4

Answer 135

4	6	8	2	5	3	9	7	1
5	9	2	7	6	1	8	3	4
3	7	1	9	4	8	2	5	6
7	4	5	3	1	2	6	9	8
2	3	6	8	9	4	5	1	7
8	1	9	5	7	6	3	4	2
6	8	7	1	3	5	4	2	9
1	5	4	6	2	9	7	8	3
9	2	3	4	8	7	1	6	5

Answer 136

1	8	4	2	5	3	7	9	6
5	2	7	6	9	4	8	3	1
9	6	3	1	7	8	4	2	5
7	4	6	3	2	9	5	1	8
8	9	2	7	1	5	3	6	4
3	5	1	8	4	6	9	7	2
2	1	5	9	8	7	6	4	3
6	7	8	4	3	2	1	5	9
4	3	9	5	6	1	2	8	7

Answer 137

9	4	5	1	7	2	6	3	8
3	7	2	4	6	8	5	9	1
6	8	1	9	5	3	7	2	4
4	2	7	8	3	9	1	6	5
8	3	6	5	2	1	4	7	9
1	5	9	7	4	6	3	8	2
5	9	8	3	1	7	2	4	6
7	6	4	2	9	5	8	1	3
2	1	3	6	8	4	9	5	7

Answer 138

8	7	1	6	2	4	3	5	9
6	2	3	9	8	5	7	4	1
4	9	5	3	1	7	6	2	8
9	4	7	5	6	2	8	1	3
3	5	8	4	7	1	9	6	2
1	6	2	8	9	3	5	7	4
2	8	9	7	4	6	1	3	5
7	3	4	1	5	9	2	8	6
5	1	6	2	3	8	4	9	7

SUD 정답 OKU

Answer 139

2	5	1	6	3	7	4	9	8
3	7	6	8	4	9	2	1	5
9	8	4	5	2	1	3	7	6
4	6	3	9	5	8	7	2	1
8	2	9	1	7	3	5	6	4
5	1	7	4	6	2	8	3	9
6	3	8	7	9	5	1	4	2
7	9	5	2	1	4	6	8	3
1	4	2	3	8	6	9	5	7

Answer 140

5	7	1	2	4	3	9	6	8
6	4	8	7	5	9	2	1	3
3	9	2	1	8	6	5	4	7
2	8	6	4	9	5	7	3	1
7	5	9	8	3	1	4	2	6
4	1	3	6	2	7	8	9	5
9	2	5	3	6	8	1	7	4
1	3	4	5	7	2	6	8	9
8	6	7	9	1	4	3	5	2

Answer 141

1	3	7	2	5	9	8	4	6
8	4	6	3	1	7	9	2	5
2	5	9	6	4	8	1	7	3
3	8	1	7	6	4	2	5	9
4	9	5	8	2	1	3	6	7
7	6	2	5	9	3	4	1	8
6	7	4	9	8	2	5	3	1
9	2	3	1	7	5	6	8	4
5	1	8	4	3	6	7	9	2

Answer 142

5	9	4	8	1	2	6	7	3
1	2	6	9	3	7	4	5	8
7	3	8	4	6	5	9	2	1
8	7	3	6	5	4	1	9	2
2	4	9	7	8	1	3	6	5
6	1	5	3	2	9	7	8	4
9	5	7	2	4	3	8	1	6
4	6	1	5	9	8	2	3	7
3	8	2	1	7	6	5	4	9

Answer 143

7	3	9	8	4	1	2	6	5
6	8	4	5	7	2	1	3	9
2	1	5	3	6	9	8	7	4
5	6	2	1	3	4	7	9	8
4	9	3	7	8	5	6	1	2
8	7	1	2	9	6	4	5	3
9	4	8	6	5	7	3	2	1
1	5	6	4	2	3	9	8	7
3	2	7	9	1	8	5	4	6

Answer 144

1	8	3	9	7	2	4	5	6
6	7	4	1	5	8	9	3	2
9	5	2	3	6	4	1	7	8
5	9	1	6	2	7	3	8	4
7	4	6	5	8	3	2	9	1
3	2	8	4	1	9	5	6	7
4	1	9	8	3	6	7	2	5
8	3	7	2	4	5	6	1	9
2	6	5	7	9	1	8	4	3

SUD정답OKU

Answer 145

5	2	4	8	6	3	7	9	1
3	6	9	1	7	4	5	8	2
1	8	7	2	5	9	6	3	4
6	5	3	9	4	1	8	2	7
9	7	1	5	8	2	4	6	3
2	4	8	6	3	7	9	1	5
4	9	5	3	1	6	2	7	8
8	3	2	7	9	5	1	4	6
7	1	6	4	2	8	3	5	9

Answer 146

6	8	7	3	9	1	2	4	5
2	4	1	5	6	7	9	3	8
5	9	3	8	4	2	7	6	1
9	6	4	7	2	5	8	1	3
3	7	5	9	1	8	4	2	6
1	2	8	6	3	4	5	7	9
4	3	6	2	5	9	1	8	7
8	1	9	4	7	3	6	5	2
7	5	2	1	8	6	3	9	4

Answer 147

9	4	5	3	8	1	6	2	7
2	7	8	9	5	6	3	4	1
6	3	1	2	7	4	5	9	8
5	9	3	8	6	2	1	7	4
1	2	6	7	4	3	8	5	9
7	8	4	1	9	5	2	3	6
4	5	2	6	1	7	9	8	3
3	6	9	4	2	8	7	1	5
8	1	7	5	3	9	4	6	2

Answer 148

9	3	7	8	6	1	2	4	5
8	4	1	2	9	5	3	6	7
6	2	5	7	3	4	1	8	9
1	9	3	6	5	7	4	2	8
2	7	4	1	8	9	6	5	3
5	6	8	3	4	2	7	9	1
3	1	9	5	2	6	8	7	4
4	8	2	9	7	3	5	1	6
7	5	6	4	1	8	9	3	2

Answer 149

7	8	2	6	9	1	5	3	4
9	1	5	7	3	4	6	8	2
3	6	4	2	8	5	7	1	9
6	9	1	5	7	8	2	4	3
2	7	8	4	6	3	9	5	1
5	4	3	1	2	9	8	7	6
4	3	7	9	5	6	1	2	8
1	2	6	8	4	7	3	9	5
8	5	9	3	1	2	4	6	7

Answer 150

7	1	8	4	6	9	2	3	5
9	5	6	2	3	8	1	7	4
3	2	4	1	7	5	9	6	8
6	8	1	3	9	2	4	5	7
5	7	9	8	4	6	3	2	1
2	4	3	5	1	7	8	9	6
1	3	2	7	5	4	6	8	9
4	9	5	6	8	3	7	1	2
8	6	7	9	2	1	5	4	3

SUD정답OKU

Answer 151

5	9	7	2	4	3	6	1	8
6	2	3	9	1	8	7	5	4
8	1	4	5	7	6	2	9	3
4	6	9	1	3	7	5	8	2
1	7	5	8	6	2	3	4	9
2	3	8	4	9	5	1	7	6
3	4	1	7	2	9	8	6	5
9	8	2	6	5	1	4	3	7
7	5	6	3	8	4	9	2	1

Answer 152

9	6	4	2	5	7	8	1	3
2	3	8	6	1	9	7	4	5
1	7	5	8	4	3	9	6	2
7	5	2	1	6	8	3	9	4
4	8	6	3	9	2	5	7	1
3	1	9	5	7	4	2	8	6
6	9	1	7	2	5	4	3	8
8	2	7	4	3	6	1	5	9
5	4	3	9	8	1	6	2	7

Answer 153

1	9	2	6	5	4	8	3	7
5	4	8	9	7	3	1	2	6
6	3	7	8	1	2	4	9	5
9	7	5	2	8	1	6	4	3
8	6	1	3	4	9	5	7	2
4	2	3	7	6	5	9	8	1
3	5	9	4	2	6	7	1	8
2	8	6	1	9	7	3	5	4
7	1	4	5	3	8	2	6	9

Answer 154

9	6	2	4	7	1	8	3	5
5	7	8	6	3	9	1	2	4
3	4	1	8	2	5	7	9	6
7	2	4	5	9	3	6	8	1
1	9	5	2	8	6	4	7	3
6	8	3	1	4	7	9	5	2
4	3	7	9	1	2	5	6	8
2	1	6	7	5	8	3	4	9
8	5	9	3	6	4	2	1	7

Answer 155

1	6	2	8	7	9	3	4	5
3	4	8	5	2	6	7	1	9
7	5	9	4	1	3	8	2	6
9	7	1	2	6	5	4	3	8
5	2	4	7	3	8	6	9	1
6	8	3	1	9	4	2	5	7
8	9	7	3	5	2	1	6	4
4	3	6	9	8	1	5	7	2
2	1	5	6	4	7	9	8	3

Answer 156

5	9	4	8	1	2	6	7	3
1	2	6	9	3	7	4	5	8
7	3	8	4	6	5	9	2	1
8	7	3	6	5	4	1	9	2
2	4	9	7	8	1	3	6	5
6	1	5	3	2	9	7	8	4
9	5	7	2	4	3	8	1	6
4	6	1	5	9	8	2	3	7
3	8	2	1	7	6	5	4	9

SUD정답OKU

Answer 157

5	6	9	1	8	7	4	3	2
2	3	1	9	4	6	7	5	8
4	8	7	2	3	5	1	6	9
9	1	5	8	7	4	3	2	6
8	4	2	3	6	9	5	7	1
3	7	6	5	1	2	9	8	4
6	9	4	7	2	3	8	1	5
7	2	8	4	5	1	6	9	3
1	5	3	6	9	8	2	4	7

Answer 158

6	4	8	7	3	9	1	2	5
9	2	3	1	8	5	4	7	6
7	5	1	4	6	2	3	8	9
2	8	7	3	4	6	9	5	1
4	9	5	2	1	7	6	3	8
3	1	6	5	9	8	7	4	2
8	3	9	6	5	4	2	1	7
5	7	4	9	2	1	8	6	3
1	6	2	8	7	3	5	9	4

Answer 159

6	9	3	2	8	4	1	5	7
8	1	2	7	5	6	9	4	3
7	4	5	1	9	3	6	2	8
9	2	7	6	1	8	5	3	4
4	8	6	3	2	5	7	1	9
5	3	1	4	7	9	2	8	6
2	7	9	8	3	1	4	6	5
3	5	4	9	6	2	8	7	1
1	6	8	5	4	7	3	9	2

Answer 160

6	9	2	5	7	4	3	1	8
7	3	1	8	2	6	4	9	5
5	4	8	3	1	9	6	7	2
2	1	4	9	6	5	7	8	3
3	6	7	2	8	1	9	5	4
9	8	5	7	4	3	1	2	6
4	7	6	1	5	8	2	3	9
8	2	9	4	3	7	5	6	1
1	5	3	6	9	2	8	4	7

Answer 161

4	2	5	9	6	8	7	1	3
6	7	8	3	1	5	9	4	2
1	9	3	7	4	2	5	8	6
9	3	6	8	7	4	2	5	1
5	8	2	6	3	1	4	7	9
7	1	4	2	5	9	3	6	8
8	4	1	5	2	3	6	9	7
3	5	7	1	9	6	8	2	4
2	6	9	4	8	7	1	3	5

Answer 162

3	2	1	5	6	4	9	7	8
5	6	9	8	3	7	4	2	1
4	7	8	9	1	2	5	3	6
8	9	5	4	2	3	1	6	7
1	4	7	6	5	8	3	9	2
2	3	6	7	9	1	8	4	5
7	8	2	3	4	5	6	1	9
6	5	4	1	7	9	2	8	3
9	1	3	2	8	6	7	5	4

SUD 정답 OKU

Answer 163

6	5	2	7	3	1	9	4	8
1	8	9	2	5	4	7	6	3
7	3	4	9	6	8	5	2	1
8	7	5	1	2	6	4	3	9
2	9	3	5	4	7	8	1	6
4	1	6	3	8	9	2	5	7
9	6	8	4	1	5	3	7	2
3	4	7	6	9	2	1	8	5
5	2	1	8	7	3	6	9	4

Answer 164

5	7	9	6	4	1	2	8	3
3	4	6	9	8	2	1	5	7
8	2	1	5	7	3	6	4	9
1	5	3	4	6	9	7	2	8
7	6	4	1	2	8	9	3	5
9	8	2	3	5	7	4	6	1
4	9	8	2	1	5	3	7	6
6	3	7	8	9	4	5	1	2
2	1	5	7	3	6	8	9	4

Answer 165

3	2	8	5	1	4	6	7	9
6	7	4	9	2	8	5	3	1
5	9	1	6	7	3	4	2	8
4	8	3	7	5	2	9	1	6
1	6	2	4	8	9	7	5	3
7	5	9	1	3	6	8	4	2
9	3	5	2	6	7	1	8	4
8	1	6	3	4	5	2	9	7
2	4	7	8	9	1	3	6	5

Answer 166

9	8	4	2	1	3	6	5	7
7	3	2	8	6	5	1	9	4
5	6	1	9	7	4	8	3	2
6	5	7	3	8	2	9	4	1
3	2	9	5	4	1	7	6	8
1	4	8	6	9	7	3	2	5
4	9	5	1	3	8	2	7	6
8	7	6	4	2	9	5	1	3
2	1	3	7	5	6	4	8	9

Answer 167

7	8	3	4	6	1	2	9	5
9	5	1	2	7	3	6	8	4
4	6	2	9	5	8	1	3	7
6	1	5	8	9	4	7	2	3
3	7	8	6	1	2	5	4	9
2	9	4	5	3	7	8	6	1
5	2	6	1	4	9	3	7	8
1	3	9	7	8	6	4	5	2
8	4	7	3	2	5	9	1	6

Answer 168

1	6	4	9	5	7	3	8	2
5	7	9	3	8	2	1	6	4
8	3	2	4	6	1	7	5	9
9	5	8	7	2	4	6	1	3
3	2	1	6	9	5	8	4	7
7	4	6	1	3	8	2	9	5
6	1	5	2	7	9	4	3	8
4	9	7	8	1	3	5	2	6
2	8	3	5	4	6	9	7	1

SUD 정답 OKU

9	1	4	7	8	6	2	3	5
5	3	7	4	2	9	6	8	1
8	6	2	1	3	5	7	9	4
7	2	3	8	6	1	5	4	9
1	5	6	3	9	4	8	7	2
4	8	9	2	5	7	3	1	6
2	7	1	6	4	8	9	5	3
6	9	8	5	1	3	4	2	7
3	4	5	9	7	2	1	6	8

5	8	9	1	7	4	3	6	2
4	3	7	6	5	2	8	1	9
2	1	6	3	8	9	5	7	4
1	7	3	5	9	6	4	2	8
8	4	5	2	1	3	6	9	7
6	9	2	8	4	7	1	5	3
3	5	1	9	2	8	7	4	6
9	6	4	7	3	1	2	8	5
7	2	8	4	6	5	9	3	1

8	7	1	6	2	4	3	5	9
6	2	3	9	8	5	7	4	1
4	9	5	3	1	7	6	2	8
9	4	7	5	6	2	8	1	3
3	5	8	4	7	1	9	6	2
1	6	2	8	9	3	5	7	4
2	8	9	7	4	6	1	3	5
7	3	4	1	5	9	2	8	6
5	1	6	2	3	8	4	9	7

7	8	9	4	2	3	5	6	1
3	4	6	9	5	1	8	7	2
1	5	2	8	7	6	3	4	9
9	6	8	2	1	7	4	5	3
4	1	3	5	6	8	2	9	7
2	7	5	3	4	9	6	1	8
5	2	7	1	3	4	9	8	6
8	3	1	6	9	5	7	2	4
6	9	4	7	8	2	1	3	5

1	2	9	4	5	7	3	6	8
3	7	6	9	8	1	2	4	5
4	5	8	6	3	2	7	1	9
8	9	7	2	4	3	1	5	6
5	3	4	1	7	6	9	8	2
6	1	2	5	9	8	4	3	7
7	8	1	3	2	5	6	9	4
2	4	3	8	6	9	5	7	1
9	6	5	7	1	4	8	2	3

6	1	8	9	2	5	4	7	3
3	7	9	6	1	4	5	8	2
5	2	4	7	3	8	9	1	6
2	4	6	1	5	3	7	9	8
8	3	7	4	6	9	1	2	5
9	5	1	2	8	7	6	3	4
4	8	2	5	9	1	3	6	7
1	6	5	3	7	2	8	4	9
7	9	3	8	4	6	2	5	1

SUD정답OKU

Answer 175

9	4	1	2	7	8	3	5	6
7	2	5	1	6	3	4	9	8
6	3	8	4	5	9	1	2	7
4	5	2	6	8	1	7	3	9
3	1	6	9	4	7	2	8	5
8	9	7	5	3	2	6	1	4
2	7	9	8	1	4	5	6	3
1	6	3	7	9	5	8	4	2
5	8	4	3	2	6	9	7	1

Answer 176

7	1	8	4	6	9	2	3	5
9	5	6	2	3	8	1	7	4
3	2	4	1	7	5	9	6	8
6	8	1	3	9	2	4	5	7
5	7	9	8	4	6	3	2	1
2	4	3	5	1	7	8	9	6
1	3	2	7	5	4	6	8	9
4	9	5	6	8	3	7	1	2
8	6	7	9	2	1	5	4	3

Answer 177

7	6	3	5	4	1	2	8	9
9	8	5	3	2	7	1	4	6
4	2	1	8	9	6	7	5	3
8	1	9	4	6	3	5	2	7
6	5	4	1	7	2	3	9	8
2	3	7	9	5	8	4	6	1
3	4	6	2	1	9	8	7	5
5	9	8	7	3	4	6	1	2
1	7	2	6	8	5	9	3	4

Answer 178

4	3	6	7	5	2	9	8	1
1	9	7	8	6	3	5	4	2
8	5	2	4	9	1	7	6	3
5	7	8	1	2	9	4	3	6
9	2	3	6	4	7	1	5	8
6	4	1	5	3	8	2	7	9
7	6	9	3	1	4	8	2	5
2	8	5	9	7	6	3	1	4
3	1	4	2	8	5	6	9	7

Answer 179

5	3	7	2	8	6	4	9	1
9	8	1	5	4	3	7	2	6
6	4	2	9	7	1	3	5	8
1	5	3	6	9	4	8	7	2
8	6	9	1	2	7	5	3	4
2	7	4	3	5	8	1	6	9
4	9	6	8	3	5	2	1	7
7	2	5	4	1	9	6	8	3
3	1	8	7	6	2	9	4	5

Answer 180

4	9	8	5	2	6	3	7	1
7	1	2	3	8	4	6	9	5
5	3	6	9	1	7	4	2	8
2	8	7	1	4	9	5	3	6
9	5	4	6	3	2	8	1	7
1	6	3	7	5	8	9	4	2
3	2	1	8	9	5	7	6	4
6	4	5	2	7	3	1	8	9
8	7	9	4	6	1	2	5	3

Answer 181

1	6	5	7	9	2	4	8	3
9	8	7	3	1	4	2	6	5
4	2	3	8	6	5	7	9	1
7	1	6	4	3	8	5	2	9
8	3	2	9	5	1	6	4	7
5	9	4	2	7	6	1	3	8
3	5	9	6	2	7	8	1	4
2	4	1	5	8	3	9	7	6
6	7	8	1	4	9	3	5	2

Answer 182

3	8	2	1	4	7	6	5	9
7	1	5	3	9	6	4	8	2
4	9	6	5	8	2	1	7	3
8	7	1	4	6	3	2	9	5
6	5	4	8	2	9	7	3	1
9	2	3	7	1	5	8	4	6
5	3	8	2	7	1	9	6	4
2	4	9	6	5	8	3	1	7
1	6	7	9	3	4	5	2	8

Answer 183

5	2	4	7	8	1	6	3	9
6	3	1	4	9	2	8	7	5
8	7	9	5	3	6	1	2	4
2	5	7	8	4	9	3	1	6
1	8	3	6	7	5	4	9	2
9	4	6	1	2	3	5	8	7
4	1	2	9	5	8	7	6	3
7	9	8	3	6	4	2	5	1
3	6	5	2	1	7	9	4	8

Answer 184

5	4	8	1	6	9	2	3	7
1	7	3	8	2	5	9	6	4
9	6	2	3	4	7	8	1	5
4	9	5	2	3	6	1	7	8
8	1	6	7	5	4	3	9	2
3	2	7	9	1	8	4	5	6
7	3	1	5	8	2	6	4	9
2	5	4	6	9	1	7	8	3
6	8	9	4	7	3	5	2	1

Answer 185

1	2	3	4	9	5	8	6	7
8	4	6	7	3	2	1	9	5
7	5	9	8	1	6	4	3	2
3	8	4	1	2	9	7	5	6
6	7	2	5	4	3	9	8	1
9	1	5	6	7	8	2	4	3
4	9	7	3	6	1	5	2	8
2	3	8	9	5	7	6	1	4
5	6	1	2	8	4	3	7	9

Answer 186

3	8	9	1	6	5	2	4	7
1	7	4	2	9	3	6	5	8
5	2	6	8	7	4	3	1	9
7	3	5	9	8	6	4	2	1
6	9	1	7	4	2	5	8	3
2	4	8	5	3	1	7	9	6
8	5	2	3	1	7	9	6	4
4	1	7	6	2	9	8	3	5
9	6	3	4	5	8	1	7	2

SUD정답OKU

9	8	4	2	1	3	6	5	7
7	3	2	8	6	5	1	9	4
5	6	1	9	7	4	8	3	2
6	5	7	3	8	2	9	4	1
3	2	9	5	4	1	7	6	8
1	4	8	6	9	7	3	2	5
4	9	5	1	3	8	2	7	6
8	7	6	4	2	9	5	1	3
2	1	3	7	5	6	4	8	9

2	4	7	5	8	3	6	9	1
8	9	3	1	4	6	7	2	5
1	6	5	9	2	7	4	8	3
6	1	4	7	3	2	8	5	9
7	5	9	4	1	8	2	3	6
3	2	8	6	5	9	1	7	4
5	8	6	2	9	1	3	4	7
4	3	1	8	7	5	9	6	2
9	7	2	3	6	4	5	1	8

9	5	6	3	4	2	1	8	7
1	3	8	6	5	7	9	4	2
2	4	7	1	9	8	3	6	5
3	1	2	9	6	5	4	7	8
7	6	9	8	3	4	5	2	1
5	8	4	2	7	1	6	9	3
8	9	3	5	2	6	7	1	4
4	2	5	7	1	9	8	3	6
6	7	1	4	8	3	2	5	9

6	1	7	4	8	9	3	2	5
9	4	2	5	3	6	7	8	1
8	5	3	1	7	2	4	6	9
3	2	9	7	1	8	5	4	6
7	6	4	3	2	5	1	9	8
5	8	1	6	9	4	2	3	7
2	3	6	9	5	1	8	7	4
1	9	8	2	4	7	6	5	3
4	7	5	8	6	3	9	1	2

8	7	3	5	9	1	4	2	6
1	4	6	2	3	7	9	5	8
2	9	5	6	8	4	1	3	7
5	1	2	8	7	9	3	6	4
6	8	7	4	2	3	5	9	1
4	3	9	1	5	6	8	7	2
3	6	1	9	4	2	7	8	5
7	2	8	3	1	5	6	4	9
9	5	4	7	6	8	2	1	3

7	2	6	1	5	8	4	3	9
5	4	8	6	3	9	7	1	2
3	9	1	2	4	7	5	8	6
8	1	3	5	7	6	9	2	4
6	7	4	9	2	3	8	5	1
2	5	9	8	1	4	6	7	3
9	6	5	3	8	1	2	4	7
4	3	2	7	9	5	1	6	8
1	8	7	4	6	2	3	9	5

SUD정답OKU

3	4	7	9	1	5	8	6	2
2	9	5	6	8	7	1	3	4
6	1	8	2	3	4	7	9	5
5	2	1	4	9	6	3	8	7
9	8	3	1	7	2	4	5	6
4	7	6	8	5	3	9	2	1
8	3	2	5	4	1	6	7	9
7	5	4	3	6	9	2	1	8
1	6	9	7	2	8	5	4	3

8	4	3	5	1	6	9	7	2
5	2	6	3	9	7	4	1	8
7	9	1	8	4	2	6	5	3
4	5	2	6	8	9	1	3	7
6	1	7	2	5	3	8	4	9
3	8	9	1	7	4	5	2	6
1	6	5	7	3	8	2	9	4
2	7	4	9	6	1	3	8	5
9	3	8	4	2	5	7	6	1

5	3	7	2	8	6	4	9	1
9	8	1	5	4	3	7	2	6
6	4	2	9	7	1	3	5	8
1	5	3	6	9	4	8	7	2
8	6	9	1	2	7	5	3	4
2	7	4	3	5	8	1	6	9
4	9	6	8	3	5	2	1	7
7	2	5	4	1	9	6	8	3
3	1	8	7	6	2	9	4	5

6	8	1	2	4	5	7	3	9
4	9	7	3	1	6	5	2	8
3	2	5	7	8	9	6	4	1
7	5	2	8	6	3	9	1	4
1	3	4	9	5	7	8	6	2
9	6	8	1	2	4	3	7	5
2	7	6	4	9	8	1	5	3
8	1	3	5	7	2	4	9	6
5	4	9	6	3	1	2	8	7

6	3	9	2	8	1	5	4	7
5	2	7	4	3	6	8	1	9
8	4	1	5	7	9	6	3	2
3	9	6	1	5	4	7	2	8
2	1	5	7	6	8	4	9	3
7	8	4	9	2	3	1	6	5
9	6	3	8	1	5	2	7	4
1	5	2	3	4	7	9	8	6
4	7	8	6	9	2	3	5	1

4	7	5	2	1	9	8	3	6
2	1	3	6	4	8	7	9	5
8	9	6	5	7	3	1	4	2
1	2	7	8	3	5	4	6	9
5	6	8	4	9	1	2	7	3
9	3	4	7	6	2	5	8	1
7	4	1	3	2	6	9	5	8
3	5	2	9	8	4	6	1	7
6	8	9	1	5	7	3	2	4

SUD 정답 OKU

Answer 199

3	5	8	9	2	4	7	6	1
1	7	2	8	3	6	9	5	4
4	9	6	1	5	7	8	2	3
7	8	4	2	9	1	5	3	6
9	1	3	6	8	5	4	7	2
6	2	5	4	7	3	1	8	9
8	4	1	7	6	2	3	9	5
2	3	7	5	1	9	6	4	8
5	6	9	3	4	8	2	1	7

Answer 200

1	7	4	8	3	5	9	6	2
9	3	8	2	6	1	5	7	4
2	6	5	7	4	9	1	3	8
8	5	2	6	9	4	7	1	3
4	9	7	3	1	8	2	5	6
6	1	3	5	2	7	8	4	9
3	8	1	4	5	2	6	9	7
7	4	9	1	8	6	3	2	5
5	2	6	9	7	3	4	8	1

Answer 201

5	7	2	4	1	9	8	6	3
6	1	4	2	8	3	9	5	7
8	9	3	6	5	7	1	2	4
1	3	6	7	2	8	4	9	5
9	4	8	5	3	1	6	7	2
7	2	5	9	4	6	3	8	1
2	8	9	1	7	4	5	3	6
3	5	1	8	6	2	7	4	9
4	6	7	3	9	5	2	1	8

Answer 202

5	9	6	2	3	1	7	8	4
1	4	8	6	7	9	2	3	5
2	7	3	4	5	8	9	6	1
7	6	2	5	1	3	8	4	9
4	1	5	9	8	2	6	7	3
8	3	9	7	4	6	1	5	2
6	8	4	1	9	5	3	2	7
9	2	7	3	6	4	5	1	8
3	5	1	8	2	7	4	9	6

Answer 203

6	8	4	7	9	5	2	3	1
3	2	5	4	8	1	7	9	6
7	1	9	6	2	3	5	4	8
1	5	2	3	6	9	4	8	7
9	4	7	2	5	8	1	6	3
8	6	3	1	4	7	9	2	5
4	7	1	9	3	6	8	5	2
2	3	8	5	7	4	6	1	9
5	9	6	8	1	2	3	7	4

Answer 204

1	4	5	8	6	2	7	9	3
6	3	8	9	4	7	1	5	2
7	2	9	3	5	1	8	6	4
8	1	3	2	9	6	4	7	5
9	6	2	4	7	5	3	1	8
4	5	7	1	3	8	6	2	9
3	9	1	6	2	4	5	8	7
2	7	6	5	8	3	9	4	1
5	8	4	7	1	9	2	3	6

SUD정답OKU

Answer 205

1	6	7	8	4	3	5	2	9
8	4	5	9	2	1	6	7	3
9	3	2	6	7	5	4	1	8
5	9	8	3	6	7	2	4	1
7	2	6	4	1	8	3	9	5
3	1	4	2	5	9	8	6	7
6	7	3	5	9	4	1	8	2
4	8	1	7	3	2	9	5	6
2	5	9	1	8	6	7	3	4

Answer 206

2	6	8	7	4	3	1	5	9
4	5	3	9	1	2	6	8	7
7	1	9	8	6	5	2	4	3
1	9	7	6	3	4	8	2	5
5	8	4	2	9	7	3	1	6
6	3	2	1	5	8	7	9	4
3	7	1	5	8	9	4	6	2
8	2	5	4	7	6	9	3	1
9	4	6	3	2	1	5	7	8

Answer 207

4	2	6	7	8	9	3	1	5
9	7	5	3	1	4	6	2	8
8	3	1	5	6	2	9	7	4
5	6	7	8	4	3	1	9	2
1	9	4	2	5	6	7	8	3
2	8	3	9	7	1	4	5	6
7	5	9	4	3	8	2	6	1
6	4	2	1	9	5	8	3	7
3	1	8	6	2	7	5	4	9

Answer 208

2	9	5	6	4	3	7	8	1
1	4	6	8	5	7	3	9	2
7	8	3	9	1	2	6	4	5
9	5	1	4	7	8	2	3	6
6	3	4	2	9	5	8	1	7
8	2	7	3	6	1	9	5	4
3	1	2	7	8	4	5	6	9
5	7	9	1	3	6	4	2	8
4	6	8	5	2	9	1	7	3

Answer 209

4	3	7	9	6	1	2	8	5
2	5	1	7	8	4	9	3	6
6	9	8	3	2	5	7	4	1
1	8	5	2	7	6	4	9	3
3	2	9	4	5	8	6	1	7
7	4	6	1	3	9	5	2	8
9	6	2	5	1	3	8	7	4
8	7	3	6	4	2	1	5	9
5	1	4	8	9	7	3	6	2

Answer 210

4	9	2	3	7	5	6	8	1
1	3	8	6	9	2	4	5	7
7	6	5	1	8	4	3	9	2
3	8	6	4	1	9	7	2	5
9	2	4	7	5	3	1	6	8
5	7	1	2	6	8	9	3	4
2	1	3	8	4	6	5	7	9
8	5	7	9	3	1	2	4	6
6	4	9	5	2	7	8	1	3

SUD정답OKU

Answer 211

4	6	9	5	7	3	8	2	1
2	7	1	9	4	8	5	3	6
5	8	3	1	2	6	7	9	4
8	5	2	3	6	7	1	4	9
9	3	7	4	5	1	6	8	2
6	1	4	8	9	2	3	5	7
1	9	8	7	3	4	2	6	5
3	2	5	6	1	9	4	7	8
7	4	6	2	8	5	9	1	3

Answer 212

8	3	7	6	4	9	1	5	2
1	9	5	7	3	2	8	4	6
2	4	6	1	8	5	3	9	7
4	7	3	8	9	1	2	6	5
9	1	2	3	5	6	4	7	8
6	5	8	2	7	4	9	3	1
5	8	9	4	1	7	6	2	3
3	2	4	5	6	8	7	1	9
7	6	1	9	2	3	5	8	4

Answer 213

1	5	9	7	8	2	3	4	6
3	2	4	9	6	5	1	7	8
6	7	8	3	4	1	9	2	5
2	8	7	4	1	3	6	5	9
4	9	3	6	5	7	8	1	2
5	6	1	8	2	9	4	3	7
8	1	5	2	3	6	7	9	4
7	3	6	5	9	4	2	8	1
9	4	2	1	7	8	5	6	3

Answer 214

1	6	8	4	9	7	2	3	5
2	5	3	8	1	6	9	4	7
9	7	4	3	5	2	8	6	1
7	9	6	2	3	5	4	1	8
8	4	5	1	7	9	3	2	6
3	2	1	6	8	4	7	5	9
4	1	9	7	6	3	5	8	2
6	3	7	5	2	8	1	9	4
5	8	2	9	4	1	6	7	3

Answer 215

9	2	1	6	5	3	4	7	8
8	4	5	2	7	9	3	6	1
3	7	6	4	1	8	2	5	9
7	5	2	9	8	1	6	4	3
1	6	8	7	3	4	5	9	2
4	9	3	5	6	2	8	1	7
6	1	7	3	2	5	9	8	4
2	8	9	1	4	6	7	3	5
5	3	4	8	9	7	1	2	6

Answer 216

5	8	3	7	4	2	1	6	9
6	9	4	3	8	1	7	5	2
2	7	1	6	5	9	4	8	3
7	5	9	2	6	8	3	1	4
3	1	6	4	9	7	8	2	5
4	2	8	1	3	5	9	7	6
9	6	5	8	7	3	2	4	1
1	3	7	5	2	4	6	9	8
8	4	2	9	1	6	5	3	7

SUD정답OKU

Answer 217

1	9	3	7	6	5	4	8	2
6	5	4	3	2	8	9	1	7
2	7	8	9	1	4	5	3	6
5	4	2	8	9	7	3	6	1
7	8	6	1	4	3	2	5	9
3	1	9	6	5	2	8	7	4
4	3	5	2	7	1	6	9	8
9	2	1	5	8	6	7	4	3
8	6	7	4	3	9	1	2	5

Answer 218

3	5	7	9	6	1	4	2	8
2	8	1	5	4	3	7	6	9
6	9	4	2	7	8	1	3	5
9	1	6	7	3	2	8	5	4
7	4	3	8	5	6	9	1	2
5	2	8	4	1	9	3	7	6
4	6	5	1	8	7	2	9	3
1	3	2	6	9	4	5	8	7
8	7	9	3	2	5	6	4	1

Answer 219

1	8	6	3	2	5	7	4	9
7	2	3	9	4	8	1	6	5
4	9	5	6	7	1	3	8	2
9	6	7	1	5	2	4	3	8
2	4	8	7	9	3	5	1	6
5	3	1	4	8	6	2	9	7
6	5	9	2	3	4	8	7	1
8	1	4	5	6	7	9	2	3
3	7	2	8	1	9	6	5	4

Answer 220

7	6	8	5	4	3	1	9	2
3	2	5	7	1	9	4	8	6
4	1	9	8	6	2	5	3	7
1	3	2	6	8	5	7	4	9
5	8	7	9	2	4	3	6	1
6	9	4	3	7	1	2	5	8
2	7	3	4	9	8	6	1	5
8	5	1	2	3	6	9	7	4
9	4	6	1	5	7	8	2	3

Answer 221

9	4	3	6	5	1	7	2	8
6	7	8	4	9	2	5	3	1
2	1	5	3	7	8	9	6	4
3	2	9	8	4	5	1	7	6
8	5	4	7	1	6	2	9	3
7	6	1	2	3	9	8	4	5
5	3	6	1	2	7	4	8	9
4	9	2	5	8	3	6	1	7
1	8	7	9	6	4	3	5	2

Answer 222

4	3	1	5	9	6	2	7	8
8	6	5	1	2	7	3	4	9
7	2	9	3	4	8	1	6	5
1	9	3	6	5	4	7	8	2
6	7	2	8	1	3	5	9	4
5	8	4	2	7	9	6	3	1
9	1	8	7	6	5	4	2	3
3	5	7	4	8	2	9	1	6
2	4	6	9	3	1	8	5	7

SUD정답OKU

Answer 223

1	7	9	5	2	8	3	4	6
5	2	6	4	7	3	9	8	1
4	8	3	6	9	1	2	7	5
9	3	5	1	4	2	8	6	7
7	6	4	9	8	5	1	2	3
2	1	8	7	3	6	4	5	9
8	9	7	3	5	4	6	1	2
3	4	1	2	6	7	5	9	8
6	5	2	8	1	9	7	3	4

Answer 224

9	7	4	5	2	8	3	1	6
6	5	2	9	1	3	4	8	7
3	8	1	4	6	7	2	9	5
5	1	6	8	3	2	9	7	4
2	9	8	7	4	6	5	3	1
4	3	7	1	5	9	8	6	2
1	6	3	2	8	4	7	5	9
8	2	9	6	7	5	1	4	3
7	4	5	3	9	1	6	2	8

Answer 225

3	6	1	9	5	2	8	4	7
9	7	8	4	1	6	3	2	5
4	2	5	3	7	8	1	6	9
5	9	3	6	2	7	4	8	1
7	1	2	8	9	4	5	3	6
6	8	4	1	3	5	9	7	2
8	3	7	5	6	9	2	1	4
1	5	6	2	4	3	7	9	8
2	4	9	7	8	1	6	5	3

Answer 226

4	9	8	7	5	2	3	6	1
3	1	2	9	4	6	5	8	7
6	5	7	1	3	8	2	9	4
7	3	9	8	6	5	1	4	2
2	8	4	3	1	7	6	5	9
1	6	5	2	9	4	8	7	3
8	4	1	5	2	9	7	3	6
5	2	6	4	7	3	9	1	8
9	7	3	6	8	1	4	2	5

Answer 227

2	1	4	6	9	8	3	5	7
7	5	9	1	3	2	6	8	4
6	8	3	7	5	4	1	2	9
5	7	6	9	2	1	8	4	3
8	3	1	4	7	5	2	9	6
4	9	2	3	8	6	7	1	5
9	2	5	8	6	3	4	7	1
1	6	7	2	4	9	5	3	8
3	4	8	5	1	7	9	6	2

Answer 228

8	2	3	7	5	4	9	1	6
9	7	6	2	1	3	8	4	5
1	5	4	8	6	9	7	2	3
6	3	5	9	7	2	4	8	1
4	9	7	6	8	1	3	5	2
2	8	1	3	4	5	6	9	7
3	4	9	5	2	6	1	7	8
5	1	8	4	3	7	2	6	9
7	6	2	1	9	8	5	3	4

SUD정답OKU

Answer 229

6	9	5	7	2	4	1	3	8
4	2	8	1	6	3	9	7	5
7	1	3	5	8	9	6	4	2
8	6	4	2	3	7	5	1	9
1	5	7	9	4	8	2	6	3
9	3	2	6	1	5	4	8	7
2	8	9	4	7	1	3	5	6
5	7	1	3	9	6	8	2	4
3	4	6	8	5	2	7	9	1

Answer 230

1	3	7	8	2	5	4	6	9
9	6	2	1	4	7	5	3	8
5	8	4	3	9	6	7	1	2
4	5	8	7	6	3	2	9	1
6	9	1	2	5	4	3	8	7
2	7	3	9	8	1	6	4	5
3	2	9	4	7	8	1	5	6
8	4	6	5	1	2	9	7	3
7	1	5	6	3	9	8	2	4

Answer 231

2	3	6	7	4	9	5	1	8
9	5	1	2	8	3	6	7	4
4	7	8	5	6	1	9	2	3
7	4	5	8	9	2	3	6	1
8	1	9	6	3	5	2	4	7
6	2	3	4	1	7	8	9	5
3	6	7	9	5	4	1	8	2
5	9	2	1	7	8	4	3	6
1	8	4	3	2	6	7	5	9

Answer 232

4	8	3	7	9	6	1	5	2
5	7	2	4	3	1	8	9	6
9	6	1	5	2	8	4	3	7
1	5	6	2	8	7	3	4	9
2	4	9	1	5	3	7	6	8
8	3	7	6	4	9	2	1	5
7	2	4	3	6	5	9	8	1
6	1	8	9	7	4	5	2	3
3	9	5	8	1	2	6	7	4

Answer 233

2	6	5	3	7	9	4	1	8
4	3	7	6	8	1	9	2	5
1	9	8	4	2	5	3	7	6
7	4	6	5	3	2	1	8	9
3	5	1	7	9	8	2	6	4
9	8	2	1	4	6	5	3	7
8	7	4	9	1	3	6	5	2
5	1	9	2	6	7	8	4	3
6	2	3	8	5	4	7	9	1

Answer 234

4	2	7	6	8	3	1	5	9
8	3	1	9	2	5	4	7	6
5	9	6	1	7	4	8	2	3
7	4	8	3	5	9	2	6	1
6	1	9	8	4	2	5	3	7
2	5	3	7	1	6	9	8	4
1	8	4	2	6	7	3	9	5
9	6	5	4	3	8	7	1	2
3	7	2	5	9	1	6	4	8

SUD정답OKU

Answer 235

9	2	3	5	4	7	6	8	1
6	1	4	8	9	3	2	5	7
8	5	7	1	6	2	9	3	4
7	9	5	2	8	4	3	1	6
3	4	2	6	7	1	8	9	5
1	8	6	3	5	9	7	4	2
5	3	9	4	2	6	1	7	8
2	7	8	9	1	5	4	6	3
4	6	1	7	3	8	5	2	9

Answer 236

5	9	6	4	1	7	8	3	2
8	3	2	9	5	6	7	4	1
4	1	7	2	3	8	5	6	9
9	8	5	6	2	3	4	1	7
3	7	1	8	4	5	9	2	6
6	2	4	1	7	9	3	5	8
2	6	3	7	8	4	1	9	5
7	4	9	5	6	1	2	8	3
1	5	8	3	9	2	6	7	4

Answer 237

9	6	2	1	7	5	8	3	4
5	8	1	4	3	6	7	2	9
3	4	7	8	2	9	1	6	5
4	1	6	2	9	8	5	7	3
2	7	9	3	5	4	6	8	1
8	3	5	6	1	7	9	4	2
7	9	3	5	6	2	4	1	8
6	2	4	9	8	1	3	5	7
1	5	8	7	4	3	2	9	6

Answer 238

9	7	1	5	2	4	3	8	6
4	8	2	7	3	6	1	9	5
6	5	3	8	1	9	4	7	2
7	2	6	9	8	1	5	3	4
1	9	8	4	5	3	2	6	7
5	3	4	6	7	2	9	1	8
8	6	9	1	4	5	7	2	3
3	1	5	2	6	7	8	4	9
2	4	7	3	9	8	6	5	1

Answer 239

4	7	2	6	5	9	3	1	8
5	1	8	4	2	3	9	7	6
6	9	3	8	1	7	4	5	2
7	8	9	2	4	5	1	6	3
2	5	4	3	6	1	7	8	9
1	3	6	9	7	8	5	2	4
3	6	5	1	8	4	2	9	7
8	4	1	7	9	2	6	3	5
9	2	7	5	3	6	8	4	1

Answer 240

6	4	9	7	3	5	8	2	1
2	5	3	8	1	4	6	7	9
1	7	8	2	6	9	3	4	5
3	2	6	9	4	1	5	8	7
5	9	1	6	8	7	4	3	2
7	8	4	3	5	2	1	9	6
9	1	5	4	2	8	7	6	3
8	6	2	1	7	3	9	5	4
4	3	7	5	9	6	2	1	8

SUD정답OKU

Answer 241

4	5	9	8	7	1	2	6	3
1	3	8	9	2	6	7	5	4
7	2	6	3	5	4	8	1	9
6	8	4	1	9	7	3	2	5
9	1	3	5	8	2	6	4	7
5	7	2	6	4	3	1	9	8
2	9	5	7	6	8	4	3	1
8	4	1	2	3	5	9	7	6
3	6	7	4	1	9	5	8	2

Answer 242

2	9	5	8	7	3	6	1	4
7	1	3	9	6	4	5	2	8
8	4	6	2	1	5	7	3	9
5	3	9	7	2	6	4	8	1
6	8	1	5	4	9	3	7	2
4	2	7	1	3	8	9	5	6
1	6	2	3	9	7	8	4	5
9	7	8	4	5	2	1	6	3
3	5	4	6	8	1	2	9	7

Answer 243

1	9	2	5	6	7	3	8	4
8	6	5	3	4	9	7	1	2
7	3	4	8	2	1	9	5	6
3	4	6	1	7	5	2	9	8
2	7	9	4	8	3	1	6	5
5	1	8	2	9	6	4	3	7
4	5	3	7	1	8	6	2	9
6	8	7	9	3	2	5	4	1
9	2	1	6	5	4	8	7	3

Answer 244

2	8	5	4	3	7	6	1	9
4	7	6	1	8	9	2	3	5
1	3	9	2	5	6	7	8	4
6	2	3	9	4	8	5	7	1
5	1	4	6	7	2	8	9	3
7	9	8	5	1	3	4	2	6
9	4	7	3	2	5	1	6	8
8	6	1	7	9	4	3	5	2
3	5	2	8	6	1	9	4	7

Answer 245

6	5	2	1	4	7	8	9	3
9	8	1	3	5	2	4	7	6
3	4	7	6	8	9	1	2	5
4	1	9	5	7	6	3	8	2
2	3	8	9	1	4	6	5	7
7	6	5	8	2	3	9	1	4
8	9	4	2	3	5	7	6	1
1	2	3	7	6	8	5	4	9
5	7	6	4	9	1	2	3	8

Answer 246

1	4	9	2	8	7	5	6	3
3	6	2	4	1	5	8	7	9
8	5	7	6	9	3	2	1	4
2	1	4	8	5	9	7	3	6
9	7	6	3	2	4	1	5	8
5	8	3	7	6	1	9	4	2
6	2	1	5	4	8	3	9	7
7	9	8	1	3	6	4	2	5
4	3	5	9	7	2	6	8	1

SUD 정답 OKU

1	9	6	2	5	3	8	4	7
8	7	3	4	6	9	5	2	1
5	4	2	1	8	7	6	9	3
3	6	4	8	2	5	7	1	9
9	8	5	3	7	1	4	6	2
7	2	1	9	4	6	3	8	5
6	5	9	7	1	8	2	3	4
4	1	8	5	3	2	9	7	6
2	3	7	6	9	4	1	5	8

9	1	2	7	4	3	5	6	8
4	8	6	2	5	9	7	1	3
7	5	3	8	1	6	2	9	4
5	3	7	4	6	2	9	8	1
6	2	8	5	9	1	3	4	7
1	9	4	3	7	8	6	5	2
8	4	9	6	2	7	1	3	5
3	7	1	9	8	5	4	2	6
2	6	5	1	3	4	8	7	9

5	8	2	6	9	7	4	3	1
3	6	7	2	1	4	5	8	9
9	4	1	8	5	3	6	7	2
1	5	8	3	6	2	7	9	4
4	3	6	5	7	9	2	1	8
2	7	9	4	8	1	3	6	5
7	9	3	1	4	5	8	2	6
6	1	4	7	2	8	9	5	3
8	2	5	9	3	6	1	4	7

4	2	5	1	3	7	8	9	6
8	1	7	9	6	4	3	5	2
9	3	6	8	2	5	7	1	4
3	5	2	4	9	1	6	7	8
7	6	8	3	5	2	1	4	9
1	9	4	7	8	6	2	3	5
5	8	9	6	1	3	4	2	7
2	7	1	5	4	8	9	6	3
6	4	3	2	7	9	5	8	1

2	6	9	3	8	5	4	7	1
8	4	5	7	6	1	9	3	2
7	1	3	4	2	9	5	8	6
4	9	8	6	1	2	7	5	3
5	3	6	9	7	4	1	2	8
1	2	7	8	5	3	6	9	4
6	8	2	1	9	7	3	4	5
3	7	1	5	4	8	2	6	9
9	5	4	2	3	6	8	1	7

5	4	7	8	6	9	1	3	2
1	9	3	2	7	4	5	8	6
8	2	6	3	5	1	7	9	4
2	7	4	6	1	8	9	5	3
9	3	1	7	4	5	2	6	8
6	5	8	9	2	3	4	1	7
4	6	9	1	3	2	8	7	5
3	1	2	5	8	7	6	4	9
7	8	5	4	9	6	3	2	1

SUD정답OKU

Answer 253

2	3	6	7	4	9	5	1	8
9	5	1	2	8	3	6	7	4
4	7	8	5	6	1	9	2	3
7	4	5	8	9	2	3	6	1
8	1	9	6	3	5	2	4	7
6	2	3	4	1	7	8	9	5
3	6	7	9	5	4	1	8	2
5	9	2	1	7	8	4	3	6
1	8	4	3	2	6	7	5	9

Answer 254

6	5	2	1	4	7	8	9	3
9	8	1	3	5	2	4	7	6
3	4	7	6	8	9	1	2	5
4	1	9	5	7	6	3	8	2
2	3	8	9	1	4	6	5	7
7	6	5	8	2	3	9	1	4
8	9	4	2	3	5	7	6	1
1	2	3	7	6	8	5	4	9
5	7	6	4	9	1	2	3	8

Answer 255

5	8	1	7	2	9	3	4	6
6	2	7	5	4	3	8	9	1
9	4	3	8	6	1	2	5	7
1	5	4	2	8	7	6	3	9
3	6	9	4	1	5	7	2	8
2	7	8	3	9	6	5	1	4
4	1	2	6	3	8	9	7	5
7	9	6	1	5	2	4	8	3
8	3	5	9	7	4	1	6	2

Answer 256

8	9	4	6	5	1	3	7	2
7	6	3	2	8	4	5	9	1
5	2	1	7	9	3	4	8	6
2	3	5	4	6	9	7	1	8
9	1	7	8	3	2	6	5	4
6	4	8	5	1	7	2	3	9
3	8	6	1	2	5	9	4	7
4	5	2	9	7	8	1	6	3
1	7	9	3	4	6	8	2	5

Answer 257

5	4	3	9	8	6	1	7	2
9	1	7	4	3	2	6	8	5
6	2	8	1	5	7	3	9	4
7	9	2	8	1	4	5	3	6
8	6	5	2	9	3	7	4	1
4	3	1	7	6	5	9	2	8
1	5	4	3	7	8	2	6	9
2	7	6	5	4	9	8	1	3
3	8	9	6	2	1	4	5	7

Answer 258

9	6	7	5	4	1	8	3	2
2	3	8	7	9	6	4	1	5
5	1	4	8	3	2	6	7	9
8	2	9	3	7	5	1	4	6
3	7	6	4	1	9	5	2	8
4	5	1	6	2	8	3	9	7
6	9	3	2	5	4	7	8	1
1	4	5	9	8	7	2	6	3
7	8	2	1	6	3	9	5	4

SUD 정답 OKU

Answer 259

2	9	1	7	5	4	6	8	3
8	4	6	2	9	3	1	5	7
5	3	7	1	6	8	4	2	9
3	7	4	6	8	5	9	1	2
1	2	5	9	4	7	3	6	8
6	8	9	3	1	2	7	4	5
7	6	3	5	2	1	8	9	4
9	5	8	4	7	6	2	3	1
4	1	2	8	3	9	5	7	6

Answer 260

5	9	3	2	6	7	1	8	4
8	6	2	9	4	1	5	3	7
7	1	4	3	8	5	6	2	9
6	8	1	7	5	2	4	9	3
4	2	7	6	9	3	8	5	1
9	3	5	8	1	4	7	6	2
2	7	8	4	3	6	9	1	5
1	4	6	5	2	9	3	7	8
3	5	9	1	7	8	2	4	6

Answer 261

5	9	2	8	3	4	7	6	1
4	7	8	1	6	2	9	3	5
1	6	3	9	5	7	8	4	2
6	5	4	3	2	8	1	7	9
2	1	7	6	4	9	3	5	8
3	8	9	5	7	1	4	2	6
9	3	5	7	1	6	2	8	4
7	4	1	2	8	5	6	9	3
8	2	6	4	9	3	5	1	7

Answer 262

1	4	5	3	2	7	8	9	6
9	6	3	8	4	1	7	5	2
7	8	2	5	9	6	4	3	1
5	7	8	1	3	9	2	6	4
2	1	9	4	6	5	3	7	8
4	3	6	2	7	8	9	1	5
6	2	1	9	8	3	5	4	7
3	5	4	7	1	2	6	8	9
8	9	7	6	5	4	1	2	3

Answer 263

6	1	2	3	4	7	8	5	9
5	3	4	8	9	1	2	6	7
9	7	8	5	6	2	1	4	3
8	4	6	7	2	9	3	1	5
7	5	9	1	3	8	4	2	6
1	2	3	6	5	4	7	9	8
2	9	7	4	8	5	6	3	1
3	8	5	2	1	6	9	7	4
4	6	1	9	7	3	5	8	2

Answer 264

4	3	7	5	9	6	1	2	8
2	9	5	1	3	8	6	4	7
6	1	8	4	7	2	5	9	3
7	8	2	9	6	3	4	5	1
9	5	4	8	1	7	3	6	2
3	6	1	2	5	4	8	7	9
1	2	6	3	4	9	7	8	5
8	4	3	7	2	5	9	1	6
5	7	9	6	8	1	2	3	4

SUD정답OKU

8	1	3	5	6	2	4	7	9
7	4	9	1	8	3	5	2	6
5	2	6	4	9	7	8	3	1
3	8	1	9	7	4	6	5	2
6	7	5	2	1	8	3	9	4
2	9	4	6	3	5	7	1	8
4	6	2	3	5	1	9	8	7
1	5	7	8	4	9	2	6	3
9	3	8	7	2	6	1	4	5

2	9	7	8	4	5	3	6	1
5	1	6	3	7	9	4	8	2
8	4	3	2	6	1	7	5	9
6	7	9	1	8	4	5	2	3
4	5	8	6	2	3	1	9	7
1	3	2	5	9	7	8	4	6
9	6	4	7	3	8	2	1	5
7	2	1	4	5	6	9	3	8
3	8	5	9	1	2	6	7	4

8	5	4	6	3	1	7	2	9
3	9	2	4	8	7	5	6	1
7	6	1	9	2	5	3	4	8
6	3	8	2	1	9	4	7	5
1	7	5	8	6	4	9	3	2
4	2	9	7	5	3	1	8	6
2	4	7	1	9	8	6	5	3
5	1	6	3	4	2	8	9	7
9	8	3	5	7	6	2	1	4

8	7	1	4	2	5	3	9	6
9	3	2	7	6	8	1	4	5
6	4	5	3	1	9	7	2	8
7	5	9	1	8	4	2	6	3
1	6	4	2	5	3	9	8	7
3	2	8	6	9	7	4	5	1
2	9	7	8	3	6	5	1	4
4	1	6	5	7	2	8	3	9
5	8	3	9	4	1	6	7	2

6	1	4	5	2	7	3	9	8
5	2	3	9	4	8	7	6	1
9	8	7	3	1	6	2	4	5
7	9	2	4	8	3	5	1	6
3	4	1	2	6	5	8	7	9
8	5	6	1	7	9	4	3	2
4	7	8	6	9	2	1	5	3
1	6	5	8	3	4	9	2	7
2	3	9	7	5	1	6	8	4

8	1	3	7	5	2	9	6	4
5	7	4	3	9	6	8	1	2
9	6	2	4	1	8	5	7	3
4	9	5	6	8	1	3	2	7
6	3	8	2	7	4	1	9	5
1	2	7	5	3	9	6	4	8
2	4	1	8	6	5	7	3	9
3	5	9	1	4	7	2	8	6
7	8	6	9	2	3	4	5	1

SUD 정답 OKU

Answer 271

1	3	5	6	9	4	7	8	2
2	8	6	3	7	1	5	9	4
9	7	4	2	5	8	3	1	6
5	6	3	9	1	7	4	2	8
8	2	1	4	6	3	9	7	5
7	4	9	8	2	5	6	3	1
3	1	7	5	4	2	8	6	9
6	5	8	1	3	9	2	4	7
4	9	2	7	8	6	1	5	3

Answer 272

1	7	4	8	6	5	9	3	2
5	3	2	7	1	9	4	6	8
9	8	6	4	3	2	1	5	7
6	9	5	2	4	1	8	7	3
8	1	3	6	5	7	2	9	4
4	2	7	3	9	8	5	1	6
7	4	1	9	2	6	3	8	5
3	9	8	5	7	4	6	2	1
2	6	5	1	8	3	7	4	9

Answer 273

5	9	2	6	1	8	4	7	3
3	1	6	7	4	5	2	8	9
4	7	8	2	3	9	5	6	1
7	2	3	5	9	1	6	4	8
8	4	9	3	6	2	1	5	7
6	5	1	8	7	4	3	9	2
1	3	4	9	5	7	8	2	6
2	6	7	4	8	3	9	1	5
9	8	5	1	2	6	7	3	4

Answer 274

6	5	2	8	4	1	7	3	9
8	3	9	5	7	2	4	6	1
1	7	4	3	9	6	5	2	8
4	1	6	9	3	8	2	7	5
9	8	5	2	6	7	3	1	4
7	2	3	4	1	5	8	9	6
3	4	1	7	5	9	6	8	2
2	6	7	1	8	4	9	5	3
5	9	8	6	2	3	1	4	7

Answer 275

9	2	5	4	8	3	7	6	1
8	3	7	6	2	1	9	5	4
1	6	4	7	9	5	3	8	2
3	9	2	5	1	6	4	7	8
7	1	6	8	4	2	5	9	3
4	5	8	3	7	9	1	2	6
2	7	1	9	3	8	6	4	5
6	4	3	2	5	7	8	1	9
5	8	9	1	6	4	2	3	7

Answer 276

5	8	4	2	6	7	9	1	3
2	6	1	3	4	9	7	8	5
7	9	3	8	5	1	2	4	6
8	4	7	9	1	5	6	3	2
9	2	6	7	3	4	1	5	8
1	3	5	6	2	8	4	7	9
3	7	2	1	8	6	5	9	4
4	1	8	5	9	2	3	6	7
6	5	9	4	7	3	8	2	1

SUD 정답 OKU

Answer 277

4	7	5	2	1	9	8	3	6
2	1	3	6	4	8	7	9	5
8	9	6	5	7	3	1	4	2
1	2	7	8	3	5	4	6	9
5	6	8	4	9	1	2	7	3
9	3	4	7	6	2	5	8	1
7	4	1	3	2	6	9	5	8
3	5	2	9	8	4	6	1	7
6	8	9	1	5	7	3	2	4

Answer 278

5	8	2	3	6	7	9	4	1
7	6	1	8	9	4	2	5	3
4	9	3	5	2	1	7	8	6
1	5	9	2	7	8	6	3	4
2	3	8	4	1	6	5	9	7
6	7	4	9	3	5	1	2	8
9	1	5	7	8	3	4	6	2
3	4	6	1	5	2	8	7	9
8	2	7	6	4	9	3	1	5

Answer 279

5	4	8	1	6	9	2	3	7
1	7	3	8	2	5	9	6	4
9	6	2	3	4	7	8	1	5
4	9	5	2	3	6	1	7	8
8	1	6	7	5	4	3	9	2
3	2	7	9	1	8	4	5	6
7	3	1	5	8	2	6	4	9
2	5	4	6	9	1	7	8	3
6	8	9	4	7	3	5	2	1

Answer 280

4	2	5	9	6	8	7	1	3
6	7	8	3	1	5	9	4	2
1	9	3	7	4	2	5	8	6
9	3	6	8	7	4	2	5	1
5	8	2	6	3	1	4	7	9
7	1	4	2	5	9	3	6	8
8	4	1	5	2	3	6	9	7
3	5	7	1	9	6	8	2	4
2	6	9	4	8	7	1	3	5

Answer 281

3	2	9	1	6	7	5	4	8
7	5	1	4	8	3	9	2	6
6	8	4	9	2	5	3	1	7
5	4	2	7	1	6	8	3	9
1	7	3	8	4	9	2	6	5
8	9	6	5	3	2	4	7	1
2	3	8	6	5	1	7	9	4
4	1	7	3	9	8	6	5	2
9	6	5	2	7	4	1	8	3

Answer 282

1	5	4	6	2	8	3	9	7
9	8	2	5	3	7	4	1	6
7	3	6	1	4	9	8	2	5
5	4	7	3	6	2	1	8	9
6	9	1	7	8	4	2	5	3
8	2	3	9	5	1	6	7	4
4	7	9	8	1	3	5	6	2
3	1	5	2	7	6	9	4	8
2	6	8	4	9	5	7	3	1

SUDOKU 정답

Answer 283

1	4	6	7	3	5	9	8	2
2	7	5	9	1	8	3	4	6
8	9	3	2	6	4	1	5	7
7	5	9	3	8	1	6	2	4
4	3	2	5	9	6	8	7	1
6	8	1	4	7	2	5	9	3
9	1	4	6	5	7	2	3	8
5	6	7	8	2	3	4	1	9
3	2	8	1	4	9	7	6	5

Answer 284

6	3	9	4	8	2	7	1	5
1	2	7	6	9	5	3	4	8
4	8	5	3	7	1	9	6	2
8	7	2	9	3	6	1	5	4
3	9	4	1	5	8	2	7	6
5	1	6	7	2	4	8	3	9
2	6	3	5	1	9	4	8	7
9	5	1	8	4	7	6	2	3
7	4	8	2	6	3	5	9	1

Answer 285

4	7	3	5	9	8	6	1	2
6	5	2	3	7	1	8	4	9
1	8	9	4	6	2	3	5	7
7	4	5	9	3	6	1	2	8
8	9	1	2	4	7	5	3	6
2	3	6	8	1	5	9	7	4
5	1	4	6	2	9	7	8	3
3	6	8	7	5	4	2	9	1
9	2	7	1	8	3	4	6	5

Answer 286

4	2	7	3	8	9	1	5	6
3	6	8	5	1	4	7	2	9
1	9	5	6	7	2	8	4	3
6	8	3	9	5	7	2	1	4
7	4	1	2	6	3	9	8	5
9	5	2	8	4	1	6	3	7
2	3	6	4	9	8	5	7	1
8	1	9	7	3	5	4	6	2
5	7	4	1	2	6	3	9	8

Answer 287

9	3	4	7	6	2	1	5	8
7	8	1	3	5	9	4	2	6
6	2	5	4	1	8	7	9	3
8	5	7	9	4	6	3	1	2
3	1	2	8	7	5	6	4	9
4	6	9	2	3	1	8	7	5
5	4	8	1	9	3	2	6	7
2	7	6	5	8	4	9	3	1
1	9	3	6	2	7	5	8	4

Answer 288

6	9	2	5	4	3	8	7	1
1	4	7	9	2	8	6	3	5
5	8	3	7	1	6	2	9	4
3	5	9	2	8	1	7	4	6
4	2	6	3	9	7	5	1	8
8	7	1	6	5	4	3	2	9
2	1	4	8	7	5	9	6	3
9	3	8	1	6	2	4	5	7
7	6	5	4	3	9	1	8	2

SUD 정답 OKU

4	8	9	5	7	1	2	6	3
3	7	2	4	6	8	9	1	5
6	5	1	9	2	3	4	8	7
5	4	7	6	9	2	8	3	1
2	3	8	1	5	7	6	4	9
9	1	6	3	8	4	7	5	2
7	9	3	8	4	5	1	2	6
8	6	5	2	1	9	3	7	4
1	2	4	7	3	6	5	9	8

6	5	1	2	7	8	4	9	3
9	3	8	6	1	4	5	2	7
4	7	2	9	5	3	1	6	8
3	6	5	8	9	1	7	4	2
7	1	4	5	3	2	9	8	6
2	8	9	4	6	7	3	5	1
8	4	3	1	2	9	6	7	5
1	2	6	7	4	5	8	3	9
5	9	7	3	8	6	2	1	4

1	9	6	2	5	3	8	4	7
8	7	3	4	6	9	5	2	1
5	4	2	1	8	7	6	9	3
3	6	4	8	2	5	7	1	9
9	8	5	3	7	1	4	6	2
7	2	1	9	4	6	3	8	5
6	5	9	7	1	8	2	3	4
4	1	8	5	3	2	9	7	6
2	3	7	6	9	4	1	5	8

7	6	9	5	1	8	4	3	2
4	3	1	6	9	2	5	8	7
2	5	8	4	3	7	1	6	9
1	7	3	8	4	9	6	2	5
8	2	6	3	5	1	9	7	4
9	4	5	2	7	6	3	1	8
3	1	2	9	8	5	7	4	6
6	9	4	7	2	3	8	5	1
5	8	7	1	6	4	2	9	3

1	5	3	6	4	2	7	8	9
2	4	7	5	9	8	1	3	6
6	8	9	1	7	3	4	5	2
7	6	5	2	3	4	8	9	1
9	1	4	8	5	6	2	7	3
8	3	2	7	1	9	5	6	4
4	7	6	9	8	1	3	2	5
5	9	1	3	2	7	6	4	8
3	2	8	4	6	5	9	1	7

7	6	2	8	3	5	1	4	9
5	8	9	1	4	7	6	3	2
4	3	1	2	6	9	7	5	8
1	5	8	4	9	3	2	7	6
2	7	4	6	5	8	3	9	1
3	9	6	7	1	2	5	8	4
9	2	3	5	8	6	4	1	7
8	4	7	3	2	1	9	6	5
6	1	5	9	7	4	8	2	3

SUD정답OKU

Answer 295

2	8	6	9	4	1	5	3	7
4	3	5	8	7	2	9	1	6
1	9	7	3	6	5	2	8	4
5	7	3	2	9	6	1	4	8
8	1	2	7	5	4	3	6	9
6	4	9	1	3	8	7	2	5
9	2	8	6	1	7	4	5	3
3	6	4	5	2	9	8	7	1
7	5	1	4	8	3	6	9	2

Answer 296

4	8	3	7	9	6	1	5	2
5	7	2	4	3	1	8	9	6
9	6	1	5	2	8	4	3	7
1	5	6	2	8	7	3	4	9
2	4	9	1	5	3	7	6	8
8	3	7	6	4	9	2	1	5
7	2	4	3	6	5	9	8	1
6	1	8	9	7	4	5	2	3
3	9	5	8	1	2	6	7	4

Answer 297

7	6	8	5	4	3	1	9	2
3	2	5	7	1	9	4	8	6
4	1	9	8	6	2	5	3	7
1	3	2	6	8	5	7	4	9
5	8	7	9	2	4	3	6	1
6	9	4	3	7	1	2	5	8
2	7	3	4	9	8	6	1	5
8	5	1	2	3	6	9	7	4
9	4	6	1	5	7	8	2	3

Answer 298

2	3	7	6	1	8	5	9	4
1	8	4	9	5	7	3	2	6
5	9	6	3	4	2	7	8	1
7	1	5	8	2	3	6	4	9
3	2	9	1	6	4	8	7	5
6	4	8	7	9	5	1	3	2
4	7	2	5	8	1	9	6	3
9	5	3	2	7	6	4	1	8
8	6	1	4	3	9	2	5	7

Answer 299

7	5	8	4	2	1	3	9	6
6	4	3	8	7	9	1	2	5
2	1	9	6	3	5	8	4	7
5	3	2	7	1	8	9	6	4
1	9	6	3	5	4	7	8	2
4	8	7	9	6	2	5	3	1
3	6	5	2	9	7	4	1	8
9	7	4	1	8	6	2	5	3
8	2	1	5	4	3	6	7	9

Answer 300

6	4	8	1	7	5	9	3	2
5	2	7	8	9	3	6	4	1
3	1	9	4	6	2	8	7	5
1	8	2	5	4	6	3	9	7
7	6	5	3	8	9	1	2	4
9	3	4	7	2	1	5	8	6
8	5	3	2	1	4	7	6	9
2	7	6	9	5	8	4	1	3
4	9	1	6	3	7	2	5	8

SUD정답OKU

Answer 301

5	7	4	6	9	8	2	1	3
6	8	1	2	3	5	7	4	9
2	9	3	7	4	1	8	6	5
1	6	7	5	2	4	3	9	8
9	4	2	8	6	3	1	5	7
8	3	5	9	1	7	6	2	4
3	5	9	1	8	2	4	7	6
4	2	6	3	7	9	5	8	1
7	1	8	4	5	6	9	3	2

Answer 302

3	6	8	4	5	1	7	2	9
5	2	1	9	6	7	4	3	8
4	9	7	8	3	2	1	6	5
7	1	4	5	9	6	3	8	2
8	5	2	3	7	4	9	1	6
9	3	6	1	2	8	5	7	4
1	7	5	6	8	9	2	4	3
2	8	3	7	4	5	6	9	1
6	4	9	2	1	3	8	5	7

Answer 303

4	1	7	3	2	8	6	9	5
3	5	9	6	4	1	2	7	8
8	6	2	9	7	5	4	1	3
5	7	6	1	9	4	3	8	2
1	8	3	5	6	2	7	4	9
2	9	4	7	8	3	5	6	1
7	2	5	8	1	6	9	3	4
6	3	8	4	5	9	1	2	7
9	4	1	2	3	7	8	5	6

Answer 304

7	6	9	4	5	1	8	3	2
8	2	1	6	3	7	5	4	9
5	3	4	2	9	8	6	7	1
9	7	5	8	1	4	2	6	3
2	4	6	5	7	3	1	9	8
1	8	3	9	2	6	4	5	7
3	9	8	1	4	5	7	2	6
6	5	7	3	8	2	9	1	4
4	1	2	7	6	9	3	8	5

Answer 305

1	8	2	4	9	3	7	6	5
5	9	4	6	2	7	3	1	8
7	6	3	1	8	5	2	9	4
8	7	1	3	5	2	6	4	9
9	2	6	7	4	8	5	3	1
3	4	5	9	1	6	8	2	7
2	1	7	5	6	9	4	8	3
6	5	9	8	3	4	1	7	2
4	3	8	2	7	1	9	5	6

Answer 306

5	2	6	3	8	9	4	7	1
8	4	9	7	1	6	2	5	3
1	3	7	2	4	5	6	9	8
3	1	4	8	5	2	7	6	9
6	7	2	4	9	3	1	8	5
9	5	8	1	6	7	3	2	4
7	9	5	6	3	1	8	4	2
4	6	1	5	2	8	9	3	7
2	8	3	9	7	4	5	1	6

SUD정답OKU

Answer 307

6	8	4	9	5	3	7	1	2
9	2	1	4	6	7	8	3	5
7	5	3	8	1	2	9	6	4
8	6	7	3	2	4	5	9	1
1	3	9	7	8	5	2	4	6
5	4	2	1	9	6	3	7	8
2	7	6	5	3	1	4	8	9
3	1	8	2	4	9	6	5	7
4	9	5	6	7	8	1	2	3

Answer 308

2	6	9	3	8	4	1	7	5
3	5	7	1	2	9	4	6	8
4	8	1	5	6	7	3	9	2
9	3	5	8	4	1	6	2	7
8	1	2	6	7	3	9	5	4
6	7	4	2	9	5	8	1	3
1	2	6	7	3	8	5	4	9
5	9	3	4	1	2	7	8	6
7	4	8	9	5	6	2	3	1

Answer 309

5	2	9	4	6	8	1	3	7
7	3	4	9	2	1	5	6	8
1	6	8	3	7	5	2	9	4
3	1	7	2	8	4	9	5	6
9	8	6	5	3	7	4	2	1
2	4	5	1	9	6	7	8	3
8	5	3	7	1	9	6	4	2
6	9	1	8	4	2	3	7	5
4	7	2	6	5	3	8	1	9

Answer 310

7	8	2	9	1	3	5	6	4
3	6	5	8	4	2	1	7	9
4	9	1	6	5	7	2	8	3
1	5	4	7	9	8	3	2	6
9	2	3	4	6	5	8	1	7
6	7	8	2	3	1	9	4	5
8	3	7	5	2	6	4	9	1
2	1	9	3	7	4	6	5	8
5	4	6	1	8	9	7	3	2

Answer 311

6	5	2	8	4	1	7	3	9
8	3	9	5	7	2	4	6	1
1	7	4	3	9	6	5	2	8
4	1	6	9	3	8	2	7	5
9	8	5	2	6	7	3	1	4
7	2	3	4	1	5	8	9	6
3	4	1	7	5	9	6	8	2
2	6	7	1	8	4	9	5	3
5	9	8	6	2	3	1	4	7

Answer 312

2	1	6	4	7	3	5	8	9
4	5	7	9	8	1	3	2	6
9	8	3	5	6	2	4	1	7
1	6	2	3	5	9	7	4	8
5	4	9	8	1	7	2	6	3
3	7	8	2	4	6	1	9	5
7	3	1	6	9	4	8	5	2
6	2	5	1	3	8	9	7	4
8	9	4	7	2	5	6	3	1

SUD정답OKU

Answer 313

8	3	4	9	7	2	5	6	1
5	1	7	8	4	6	3	9	2
2	9	6	5	3	1	4	7	8
4	5	8	6	1	9	2	3	7
9	7	2	3	8	4	6	1	5
3	6	1	7	2	5	9	8	4
1	8	5	2	6	3	7	4	9
6	4	9	1	5	7	8	2	3
7	2	3	4	9	8	1	5	6

Answer 314

4	3	2	9	1	8	7	5	6
1	7	8	4	5	6	3	2	9
5	6	9	2	7	3	8	4	1
9	4	6	5	3	1	2	7	8
7	2	1	8	9	4	5	6	3
8	5	3	6	2	7	1	9	4
3	1	5	7	4	9	6	8	2
6	9	7	1	8	2	4	3	5
2	8	4	3	6	5	9	1	7

Answer 315

6	5	1	2	7	8	4	9	3
9	3	8	6	1	4	5	2	7
4	7	2	9	5	3	1	6	8
3	6	5	8	9	1	7	4	2
7	1	4	5	3	2	9	8	6
2	8	9	4	6	7	3	5	1
8	4	3	1	2	9	6	7	5
1	2	6	7	4	5	8	3	9
5	9	7	3	8	6	2	1	4

Answer 316

3	9	2	1	7	8	5	4	6
4	8	5	6	2	3	7	1	9
6	7	1	4	5	9	3	2	8
8	1	7	2	9	4	6	3	5
2	6	4	5	3	7	8	9	1
5	3	9	8	6	1	4	7	2
7	5	6	3	1	2	9	8	4
9	2	8	7	4	5	1	6	3
1	4	3	9	8	6	2	5	7

Answer 317

7	1	5	8	4	9	6	3	2
6	4	2	3	1	7	8	5	9
8	9	3	5	6	2	1	4	7
3	6	7	4	5	1	2	9	8
1	2	8	9	3	6	4	7	5
4	5	9	7	2	8	3	1	6
2	3	6	1	7	5	9	8	4
5	8	1	2	9	4	7	6	3
9	7	4	6	8	3	5	2	1

Answer 318

9	8	7	6	2	3	1	5	4
2	4	1	9	8	5	6	3	7
5	3	6	4	7	1	8	2	9
7	2	4	3	6	9	5	8	1
6	9	5	1	4	8	2	7	3
3	1	8	7	5	2	4	9	6
8	7	3	5	1	4	9	6	2
1	5	9	2	3	6	7	4	8
4	6	2	8	9	7	3	1	5

SUD 정답 OKU

Answer 319

7	1	9	2	6	5	8	3	4
3	2	8	4	7	9	5	6	1
6	4	5	3	1	8	2	9	7
8	6	1	7	5	2	3	4	9
2	9	7	8	3	4	6	1	5
5	3	4	6	9	1	7	8	2
1	7	3	5	4	6	9	2	8
4	5	2	9	8	3	1	7	6
9	8	6	1	2	7	4	5	3

Answer 320

6	8	4	7	9	5	2	3	1
3	2	5	4	8	1	7	9	6
7	1	9	6	2	3	5	4	8
1	5	2	3	6	9	4	8	7
9	4	7	2	5	8	1	6	3
8	6	3	1	4	7	9	2	5
4	7	1	9	3	6	8	5	2
2	3	8	5	7	4	6	1	9
5	9	6	8	1	2	3	7	4

Answer 321

4	2	5	7	9	3	6	8	1
6	7	3	8	5	1	9	2	4
1	8	9	2	4	6	3	5	7
9	6	4	5	8	7	1	3	2
3	1	8	6	2	4	7	9	5
2	5	7	3	1	9	4	6	8
7	3	2	4	6	8	5	1	9
5	4	1	9	3	2	8	7	6
8	9	6	1	7	5	2	4	3

Answer 322

6	5	2	4	3	8	9	7	1
8	7	3	6	9	1	5	2	4
9	1	4	5	2	7	6	8	3
5	3	8	9	1	2	4	6	7
1	6	7	8	4	5	2	3	9
2	4	9	3	7	6	8	1	5
4	2	5	1	6	3	7	9	8
3	8	6	7	5	9	1	4	2
7	9	1	2	8	4	3	5	6

Answer 323

8	6	3	1	2	5	9	7	4
5	7	4	3	6	9	1	8	2
9	1	2	4	7	8	6	5	3
3	5	9	8	1	6	2	4	7
4	2	1	7	9	3	5	6	8
7	8	6	2	5	4	3	1	9
6	9	8	5	3	7	4	2	1
1	3	7	6	4	2	8	9	5
2	4	5	9	8	1	7	3	6

Answer 324

2	6	8	7	4	3	1	5	9
4	5	3	9	1	2	6	8	7
7	1	9	8	6	5	2	4	3
1	9	7	6	3	4	8	2	5
5	8	4	2	9	7	3	1	6
6	3	2	1	5	8	7	9	4
3	7	1	5	8	9	4	6	2
8	2	5	4	7	6	9	3	1
9	4	6	3	2	1	5	7	8

SUD 정답 OKU

Answer 325

1	5	9	2	8	3	6	7	4
2	7	6	9	4	1	5	3	8
4	8	3	7	6	5	2	9	1
6	1	8	5	7	4	3	2	9
7	9	4	3	1	2	8	6	5
5	3	2	6	9	8	4	1	7
8	2	5	1	3	9	7	4	6
3	6	1	4	5	7	9	8	2
9	4	7	8	2	6	1	5	3

Answer 326

4	2	7	6	8	3	1	5	9
8	3	1	9	2	5	4	7	6
5	9	6	1	7	4	8	2	3
7	4	8	3	5	9	2	6	1
6	1	9	8	4	2	5	3	7
2	5	3	7	1	6	9	8	4
1	8	4	2	6	7	3	9	5
9	6	5	4	3	8	7	1	2
3	7	2	5	9	1	6	4	8

Answer 327

5	9	6	2	3	1	7	8	4
1	4	8	6	7	9	2	3	5
2	7	3	4	5	8	9	6	1
7	6	2	5	1	3	8	4	9
4	1	5	9	8	2	6	7	3
8	3	9	7	4	6	1	5	2
6	8	4	1	9	5	3	2	7
9	2	7	3	6	4	5	1	8
3	5	1	8	2	7	4	9	6

Answer 328

6	4	7	3	5	1	9	8	2
2	3	1	7	9	8	6	4	5
9	8	5	6	4	2	1	7	3
3	2	6	9	8	4	5	1	7
5	1	9	2	6	7	4	3	8
4	7	8	5	1	3	2	9	6
8	5	3	4	2	9	7	6	1
1	9	2	8	7	6	3	5	4
7	6	4	1	3	5	8	2	9

Answer 329

7	3	4	2	8	5	6	9	1
5	8	1	6	7	9	3	4	2
9	6	2	1	3	4	7	5	8
6	1	7	5	4	3	2	8	9
3	5	8	9	6	2	4	1	7
2	4	9	8	1	7	5	3	6
8	2	3	4	9	6	1	7	5
4	9	5	7	2	1	8	6	3
1	7	6	3	5	8	9	2	4

Answer 330

7	8	3	4	6	1	2	9	5
9	5	1	2	7	3	6	8	4
4	6	2	9	5	8	1	3	7
6	1	5	8	9	4	7	2	3
3	7	8	6	1	2	5	4	9
2	9	4	5	3	7	8	6	1
5	2	6	1	4	9	3	7	8
1	3	9	7	8	6	4	5	2
8	4	7	3	2	5	9	1	6

SUD 정답 OKU

4	7	9	8	2	6	5	3	1
1	8	5	7	3	4	9	6	2
6	3	2	1	9	5	4	7	8
9	2	3	6	5	1	7	8	4
7	1	4	2	8	3	6	9	5
8	5	6	9	4	7	1	2	3
2	6	7	5	1	8	3	4	9
5	4	8	3	7	9	2	1	6
3	9	1	4	6	2	8	5	7

9	4	5	3	8	1	6	2	7
2	7	8	9	5	6	3	4	1
6	3	1	2	7	4	5	9	8
5	9	3	8	6	2	1	7	4
1	2	6	7	4	3	8	5	9
7	8	4	1	9	5	2	3	6
4	5	2	6	1	7	9	8	3
3	6	9	4	2	8	7	1	5
8	1	7	5	3	9	4	6	2

5	4	6	3	1	9	8	7	2
1	3	2	7	6	8	9	4	5
8	7	9	2	5	4	1	6	3
4	5	8	6	7	2	3	9	1
9	1	7	8	4	3	5	2	6
6	2	3	1	9	5	7	8	4
3	8	1	4	2	7	6	5	9
2	6	5	9	8	1	4	3	7
7	9	4	5	3	6	2	1	8

4	7	1	8	2	3	6	5	9
8	2	9	5	1	6	4	7	3
5	3	6	7	9	4	2	8	1
3	9	5	6	7	8	1	4	2
7	8	4	2	3	1	9	6	5
1	6	2	4	5	9	8	3	7
2	5	8	1	6	7	3	9	4
9	4	7	3	8	2	5	1	6
6	1	3	9	4	5	7	2	8

1	9	2	6	5	4	8	3	7
5	4	8	9	7	3	1	2	6
6	3	7	8	1	2	4	9	5
9	7	5	2	8	1	6	4	3
8	6	1	3	4	9	5	7	2
4	2	3	7	6	5	9	8	1
3	5	9	4	2	6	7	1	8
2	8	6	1	9	7	3	5	4
7	1	4	5	3	8	2	6	9

9	6	2	1	7	5	8	3	4
5	8	1	4	3	6	7	2	9
3	4	7	8	2	9	1	6	5
4	1	6	2	9	8	5	7	3
2	7	9	3	5	4	6	8	1
8	3	5	6	1	7	9	4	2
7	9	3	5	6	2	4	1	8
6	2	4	9	8	1	3	5	7
1	5	8	7	4	3	2	9	6

SUD정답OKU

Answer 337

4	7	9	3	2	8	6	5	1
6	5	3	1	4	9	2	7	8
2	8	1	6	7	5	9	3	4
5	9	8	2	6	4	3	1	7
7	4	6	5	3	1	8	9	2
3	1	2	8	9	7	4	6	5
9	6	4	7	5	2	1	8	3
1	2	7	9	8	3	5	4	6
8	3	5	4	1	6	7	2	9

Answer 338

5	7	8	2	9	4	6	3	1
3	1	4	6	5	7	9	8	2
9	2	6	3	8	1	4	7	5
2	8	9	5	1	6	3	4	7
4	3	1	7	2	8	5	9	6
6	5	7	4	3	9	2	1	8
1	6	2	9	7	3	8	5	4
8	9	5	1	4	2	7	6	3
7	4	3	8	6	5	1	2	9

Answer 339

9	5	3	1	7	2	4	8	6
1	7	6	4	3	8	5	9	2
8	2	4	5	6	9	1	3	7
5	8	1	2	4	6	9	7	3
3	9	2	7	5	1	8	6	4
4	6	7	8	9	3	2	5	1
6	3	5	9	2	4	7	1	8
7	4	8	6	1	5	3	2	9
2	1	9	3	8	7	6	4	5

Answer 340

7	1	5	8	4	9	6	3	2
6	4	2	3	1	7	8	5	9
8	9	3	5	6	2	1	4	7
3	6	7	4	5	1	2	9	8
1	2	8	9	3	6	4	7	5
4	5	9	7	2	8	3	1	6
2	3	6	1	7	5	9	8	4
5	8	1	2	9	4	7	6	3
9	7	4	6	8	3	5	2	1

Answer 341

9	3	2	6	4	8	7	5	1
1	6	5	9	3	7	2	8	4
4	8	7	2	1	5	9	3	6
6	7	9	4	8	1	3	2	5
8	1	3	5	2	6	4	7	9
2	5	4	7	9	3	6	1	8
5	9	8	3	7	4	1	6	2
3	4	6	1	5	2	8	9	7
7	2	1	8	6	9	5	4	3

Answer 342

4	9	3	6	5	7	1	2	8
8	2	7	9	1	4	5	6	3
6	5	1	8	3	2	9	7	4
2	7	6	5	9	8	3	4	1
5	3	4	2	7	1	8	9	6
1	8	9	3	4	6	2	5	7
3	4	8	7	2	5	6	1	9
7	6	5	1	8	9	4	3	2
9	1	2	4	6	3	7	8	5

SUD정답OKU

443

Answer 343

1	4	6	8	3	5	7	9	2
3	5	7	2	6	9	4	1	8
8	2	9	4	7	1	3	6	5
4	3	8	1	9	2	5	7	6
2	7	1	6	5	4	8	3	9
9	6	5	7	8	3	2	4	1
7	8	3	5	1	6	9	2	4
6	9	4	3	2	8	1	5	7
5	1	2	9	4	7	6	8	3

Answer 344

4	6	7	1	2	8	3	5	9
1	9	2	5	3	4	7	6	8
3	8	5	9	7	6	2	4	1
9	2	1	8	5	7	6	3	4
7	4	3	6	9	1	8	2	5
8	5	6	2	4	3	9	1	7
6	1	9	4	8	2	5	7	3
2	3	8	7	1	5	4	9	6
5	7	4	3	6	9	1	8	2

Answer 345

7	2	1	9	5	4	3	6	8
8	4	9	6	1	3	7	2	5
3	6	5	2	7	8	9	4	1
2	8	6	5	3	9	4	1	7
1	3	7	4	8	6	5	9	2
5	9	4	1	2	7	6	8	3
9	5	8	7	6	2	1	3	4
4	1	3	8	9	5	2	7	6
6	7	2	3	4	1	8	5	9

Answer 346

6	4	5	9	2	3	8	7	1
7	8	2	6	1	5	9	3	4
3	9	1	7	8	4	5	2	6
5	1	8	3	4	6	2	9	7
4	3	9	2	7	8	1	6	5
2	7	6	1	5	9	3	4	8
8	6	4	5	9	2	7	1	3
9	5	7	4	3	1	6	8	2
1	2	3	8	6	7	4	5	9

Answer 347

7	8	3	4	6	9	2	5	1
9	2	4	3	1	5	7	8	6
5	1	6	7	8	2	4	9	3
4	3	8	9	5	1	6	2	7
6	7	5	8	2	4	1	3	9
1	9	2	6	3	7	5	4	8
2	4	9	1	7	8	3	6	5
3	5	1	2	9	6	8	7	4
8	6	7	5	4	3	9	1	2

Answer 348

8	6	3	1	7	2	9	5	4
4	9	7	6	5	3	2	8	1
5	1	2	4	9	8	3	7	6
2	7	5	3	8	4	1	6	9
1	4	9	5	2	6	7	3	8
6	3	8	9	1	7	4	2	5
3	8	6	7	4	1	5	9	2
7	5	4	2	6	9	8	1	3
9	2	1	8	3	5	6	4	7

SUD정답OKU

Answer 349

6	4	1	9	3	8	5	7	2
9	8	5	7	2	6	3	1	4
3	2	7	1	5	4	6	8	9
1	3	6	5	4	9	8	2	7
2	7	9	8	1	3	4	5	6
8	5	4	2	6	7	1	9	3
7	9	3	6	8	1	2	4	5
5	6	8	4	7	2	9	3	1
4	1	2	3	9	5	7	6	8

Answer 350

9	1	8	3	2	5	6	7	4
2	5	6	4	7	8	3	9	1
3	7	4	1	9	6	5	8	2
7	2	1	6	4	3	9	5	8
4	6	9	8	5	2	7	1	3
5	8	3	9	1	7	4	2	6
6	4	2	5	8	9	1	3	7
8	3	5	7	6	1	2	4	9
1	9	7	2	3	4	8	6	5

Answer 351

1	7	3	9	4	8	6	2	5
5	8	4	6	2	7	3	1	9
6	2	9	1	5	3	8	7	4
2	3	8	5	6	4	1	9	7
4	6	1	2	7	9	5	8	3
9	5	7	8	3	1	4	6	2
7	9	5	4	1	6	2	3	8
8	1	2	3	9	5	7	4	6
3	4	6	7	8	2	9	5	1

Answer 352

8	2	3	7	1	6	4	5	9
1	9	4	5	2	8	7	3	6
6	5	7	4	9	3	2	8	1
7	4	9	3	8	1	6	2	5
2	3	6	9	4	5	8	1	7
5	8	1	2	6	7	9	4	3
3	1	8	6	7	2	5	9	4
4	7	5	8	3	9	1	6	2
9	6	2	1	5	4	3	7	8

Answer 353

7	8	9	4	2	3	5	6	1
3	4	6	9	5	1	8	7	2
1	5	2	8	7	6	3	4	9
9	6	8	2	1	7	4	5	3
4	1	3	5	6	8	2	9	7
2	7	5	3	4	9	6	1	8
5	2	7	1	3	4	9	8	6
8	3	1	6	9	5	7	2	4
6	9	4	7	8	2	1	3	5

Answer 354

2	5	6	3	8	1	9	7	4
7	8	4	9	5	2	6	1	3
9	1	3	7	6	4	8	5	2
3	2	8	1	7	9	4	6	5
5	6	1	2	4	8	7	3	9
4	7	9	6	3	5	2	8	1
1	4	5	8	9	6	3	2	7
6	3	2	4	1	7	5	9	8
8	9	7	5	2	3	1	4	6

SUD 정답 OKU

Answer 355

7	1	6	9	4	5	3	2	8
9	3	8	6	2	1	4	5	7
5	4	2	3	7	8	6	9	1
4	9	5	7	3	2	8	1	6
6	8	7	1	5	9	2	4	3
3	2	1	8	6	4	5	7	9
8	6	4	2	1	7	9	3	5
2	7	3	5	9	6	1	8	4
1	5	9	4	8	3	7	6	2

Answer 356

1	6	5	7	9	2	4	8	3
9	8	7	3	1	4	2	6	5
4	2	3	8	6	5	7	9	1
7	1	6	4	3	8	5	2	9
8	3	2	9	5	1	6	4	7
5	9	4	2	7	6	1	3	8
3	5	9	6	2	7	8	1	4
2	4	1	5	8	3	9	7	6
6	7	8	1	4	9	3	5	2

Answer 357

1	9	3	7	4	6	2	8	5
2	8	4	9	1	5	3	7	6
5	7	6	2	3	8	1	9	4
7	1	5	3	8	4	9	6	2
3	4	9	6	2	7	8	5	1
6	2	8	5	9	1	4	3	7
4	3	7	8	6	2	5	1	9
9	6	1	4	5	3	7	2	8
8	5	2	1	7	9	6	4	3

Answer 358

6	3	1	5	9	4	7	2	8
7	4	5	2	3	8	1	6	9
2	9	8	6	1	7	3	5	4
5	2	6	9	8	1	4	7	3
3	1	4	7	2	5	9	8	6
8	7	9	3	4	6	5	1	2
4	5	2	1	6	3	8	9	7
1	6	3	8	7	9	2	4	5
9	8	7	4	5	2	6	3	1

Answer 359

2	5	3	7	8	1	9	6	4
7	6	4	3	9	5	1	2	8
8	1	9	2	4	6	7	3	5
3	2	8	9	1	4	6	5	7
5	4	7	6	3	2	8	1	9
1	9	6	5	7	8	3	4	2
9	8	2	4	6	3	5	7	1
4	3	1	8	5	7	2	9	6
6	7	5	1	2	9	4	8	3

Answer 360

4	6	8	2	5	3	9	7	1
5	9	2	7	6	1	8	3	4
3	7	1	9	4	8	2	5	6
7	4	5	3	1	2	6	9	8
2	3	6	8	9	4	5	1	7
8	1	9	5	7	6	3	4	2
6	8	7	1	3	5	4	2	9
1	5	4	6	2	9	7	8	3
9	2	3	4	8	7	1	6	5

SUD 정답 OKU

Answer 361

1	2	3	4	8	6	5	7	9
7	4	8	2	5	9	1	3	6
6	9	5	7	1	3	2	4	8
8	5	6	3	4	2	7	9	1
2	7	4	5	9	1	8	6	3
9	3	1	8	6	7	4	2	5
5	6	9	1	2	4	3	8	7
4	1	7	6	3	8	9	5	2
3	8	2	9	7	5	6	1	4

Answer 362

9	4	5	2	6	8	3	1	7
2	1	6	9	7	3	8	4	5
7	8	3	1	5	4	2	9	6
4	6	9	7	3	2	1	5	8
8	5	2	4	1	9	6	7	3
3	7	1	5	8	6	4	2	9
6	2	7	3	4	5	9	8	1
5	3	4	8	9	1	7	6	2
1	9	8	6	2	7	5	3	4

Answer 363

4	8	1	3	6	5	9	7	2
3	7	9	8	2	4	6	5	1
5	6	2	7	1	9	3	8	4
8	5	6	1	3	2	7	4	9
9	2	7	5	4	8	1	3	6
1	3	4	6	9	7	8	2	5
7	1	8	2	5	6	4	9	3
2	9	3	4	8	1	5	6	7
6	4	5	9	7	3	2	1	8

Answer 364

3	7	8	2	5	4	6	1	9
2	4	1	9	6	8	7	5	3
5	9	6	1	3	7	8	2	4
6	2	7	8	4	9	5	3	1
4	1	3	6	7	5	9	8	2
8	5	9	3	2	1	4	7	6
9	8	2	7	1	6	3	4	5
7	3	5	4	9	2	1	6	8
1	6	4	5	8	3	2	9	7

Answer 365

5	4	2	9	3	1	8	6	7
9	1	6	4	8	7	2	5	3
8	7	3	6	2	5	9	4	1
2	8	1	3	6	9	5	7	4
4	6	5	7	1	8	3	2	9
3	9	7	2	5	4	6	1	8
6	5	4	8	7	3	1	9	2
7	2	8	1	9	6	4	3	5
1	3	9	5	4	2	7	8	6

브레인 트레이닝 스도쿠 365

1쇄 발행	2020년 08월 20일
3쇄 발행	2022년 01월 03일
지은이	스도쿠 동호회
펴낸이	김왕기
편집부	원선화, 김한솔
디자인	푸른영토 디자인실
펴낸곳	**푸른e미디어**

주소	경기도 고양시 일산동구 장항동 865 코오롱레이크폴리스1차 A동 908호
전화	(대표)031-925-2327, 070-7477-0386~9 · 팩스	031-925-2328
등록번호	제2005-24호.(2005년 4월 15일)
홈페이지	www.blueterritory.com
전자우편	designkwk@me.com

ISBN 979-11-88287-05-5   14410
ⓒ스도쿠 동호회, 2020

푸른e미디어는 (주)푸른영의 임프린트입니다.